New fibers

New fibers

2nd edition

TATSUYA HONGU, B.Eng., M.Eng., Ph.D.
Toho Rayon Co Ltd, Tokyo, Japan

GLYN O. PHILLIPS, Ph.D., D.Sc., Hon.D.Sc., C.Chem., F.R.S.C.
Research Transfer Ltd, Wales, UK and
Visiting Professor of Chemistry, University of Salford, UK

WOODHEAD PUBLISHING LIMITED
Cambridge England

First published, 1990, Ellis Horwood Ltd
Second edition, 1997, Woodhead Publishing Ltd

British Library Cataloguing in Publication Data
A catalogue record for this book is available from the British Library.

ISBN 1 85573 334 X

Designed by Geoff Green
Typeset by Euroset, Alresford, Hampshire SO24 9PG, England
Printed by St Edmundsbury Press, Suffolk, England

Contents

Preface to the first edition

This new insight into the world of New Fibers is presented from a Japanese perspective. This deliberate emphasis has arisen for two reasons. First, the basis of the material was the extremely successful Japanese text by Dr Hongu, which has already sold more than 20,000 copies, and is now in its 5th edition. Secondly, the treatment illustrates the unique Japanese approach to technical developments, once the basic break-through has been achieved.

The "super-fibers" which emerged during the 1980s are undoubtedly a chemical and technological triumph. Nevertheless, the excellence of their material properties could not have been given full scope without the imaginative exploitation which has been led by the Japanese. To achieve the aesthetic qualities of natural fur, silk, wool and leather, and to greatly improve on their performance using synthetic materials, require technical mimicry and a fundamental understanding of the properties and manner in which the natural fibers were created. Nature is able to introduce particular structures which impart their own characteristic touch, smell, colour and biological function to biopolymers. This book explains how these processes were studied and subsequently imitated.

The sheer beauty and economy of nature enables complicated polymers first to be fabricated, which then control living processes. The harnessing of these secrets has now led to the availability of new biopolymers which can impart living characteristics into the inanimate. The totality of this chemical, technical, biosynthetic and biomimetic approach has now yielded the high-tech fibers, which find new applications in areas as varied as electronics, medicine, space, nuclear power, the oceans, the earth and the race for perfection in sport.

It is directed mainly at the scientist who is broad in his technical interest and the layman who seeks a glimpse of this brave new world of New Fibers. There is enough information for the specialist to follow up the product and commercial leads which may be relevant to his work, but is not meant, however, to be a technical manual. Students of fiber science will not find this information collected so coherently elsewhere, and the book could serve as a starting point in their exploration of the subject. To Western scientists and technologists, the Japanese approach might prove a revelation. Above all, we hope the reader enjoys this fascinating subject.

Finally, we would like to acknowledge and thank the persons who have assisted us in the production of this volume. Dr Kanji Kajiwara (Kyoto Institute of Technology) and Ms Machiko Takigami (Gunma University) have given invaluable support in interpreting and translating the Japanese data. Ms Linda Sneddon undertook all the typing and Stephen Williams all the art work. Together these two set out the pages [of the 1st edition] in their final form. We warmly thank our four colleagues for their complete dedication and for giving selflessly of their professional expertise.

TATSUYA HONGU
GLYN O. PHILLIPS

Preface to the second edition

We were pleased at the enthusiastic reception given to the 1st edition of *New Fibers*. It was found useful by students and practitioners in fiber science. We were surprised that the demand was so great from colleges in Asia and the Pacific Region, where English is not the first language. This was particularly true for Japan, where students used the book simultaneously to improve their knowledge about New fibers and to communicate their subject in English. The demand soon outstripped the supply, necessitating a new edition. We have taken this opportunity to revise the original and to add new information about developments since 1990, when the 1st edition was published.

Since 1990 there have been considerable changes in the nature of the fibers being produced, the production methods and in consumers' values and expectations. The quantity of fibers produced in the ASEAN countries over this period has increased, whereas the production has decreased in Japan. Nevertheless, it was Japanese fiber technology that continued to drive the new developments. As with the 1st edition we have made an effort to interpret the Japanese viewpoint to a wider audience. Since 1990, the march of High-tech fibers has continued, with an ever-increasing sub-division to meet the specialised applications, as in high-performance, high-function and high-sense fibers.

New research and development has produced fibers with high tenacity and modulus to give the Super-fibers, now used as industrial materials. The more aesthetic and comfortable ways of life have been given rise to the improved and new *Shin-gosen*, the ultra-fine fibers that can emulate the functionality and ambience of biological fibers. It is this springboard that leads on to "fibers for the next millennium", the subject of a new chapter. The new high-tech fibers have also been given a new chapter to bring the subject up to date. Where necessary, statistical information has also been updated.

The synthetic cellulosics have made a particular resurgence since 1990, and the various solvent-spun fibers of the *Lyocell* and *Tencel* families are now making a great impact on the market. They now offer the processability of synthetics along with the in-built advantages of natural cotton. Accordingly, a new chapter has been added to deal with this subject.

We hope, therefore, that the approach of the 1st edition is now extended to illustrate the dynamism of this frontier industry, pointing the way forward into the next century. It is not primarily directed to specialist fiber technologists, but aims to inform students in fiber science and others who are engaged in the fiber industry and related fields. Students in basic chemistry and physics will find it helpful to give them an insight into this fascinating subject. Above all we hope that the reader enjoys the subject, as we have done during its preparation.

Our two friends Machiko Takigami (now Dr Takigami), Takasaki Radiation Chemistry Research Establishment, Japan Atomic Energy Research Institute and Kanji Kajiwara (now Professor Kajiwara), Kyoto Institute of Technology have greatly assisted us again in preparing this 2nd edition. Both have gained further distinction in the intervening period since 1990. We wish to give them our sincere thanks and also to Stephen Williams who again has helped with the art work. They have given freely of their professional expertise, which has enabled us to communicate across the world and convey the inherent Japanese thinking that has so revolutionised this subject. Finally, we thank Patricia Morrison and Amanda Thomas of Woodhead Publishing Ltd, for ensuring that this new edition sees the light of day.

TATSUYA HONGU
GLYN O. PHILLIPS

1 Birth of the new fibers

1.1 Background

Two types of fibers are currently available to human society: natural fibers, which have existed for 4,000 years or more, and synthetic fibers, which first appeared about 100 years ago, when Count Chardonnet invented artificial silk, an achievement that had been only a human dream previously. Dr Carothers of the Du Pont Company first produced nylon in 1935, which was claimed to be "finer than spider's thread, stronger than steel and more elegant than silk". Today, synthetic fibers are not a mere alternative to natural fibers, but are new materials of high functionality and high performance which play a key role in the field of high technology. Now these new materials can be designed and produced according to the nature of their utilisation.

Macroscopically, the era of natural fibers, which existed up to the 1950s, can be identified as the first generation; the synthetic fibers such as nylon, polyester, polyacrylonitrile, etc. that appeared during the 1950s were the second generation (see Fig. 1.1). These were the chemist's copy of natural fibers, in order to replace them, and to some extent they succeeded in this purpose. However, the fiber materials of high performance in use today provide the potential for developing a new technology. These fiber materials can be classified as third generation. Fibers of high modulus and high strength can now be further produced from synthetic polymers of light weight, and are widely employed in space technology, as, for example, high-tenacity fibers of polyethylene, polyaramid and polyarylate. Although carbon fiber is in nature inorganic, it is conventionally classified as a synthetic fiber, since it is mainly produced from polyacrylonitrile (PAN).

Thus synthetic fibers of the third generation are not simply alternatives to natural fibers, as the synthetic fibers of the second generation were. The need for ultra-light fibers of high strength is increasing as high technology responds to changes in the social environment, for example, by the demands of energy

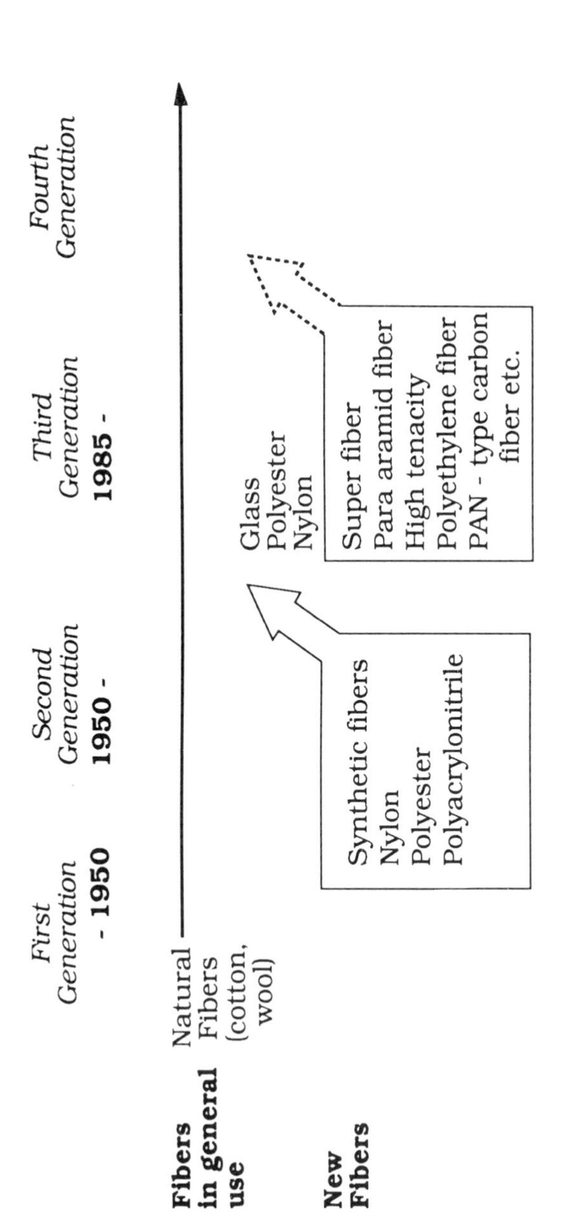

1.1 Fiber development in each generation.

conservation. Such fibers are also needed in various sports, leisure, transportation, and ocean/air/space developments, leading out from Earth to space. In future decades, metals are expected to be replaced by newly developed synthetic fibers of the third generation, which can be superior to metals with respect to strength and modulus. These fibers are already being employed as strengthening materials in composites in main wings and/or other body parts of aircraft and space shuttles.

1.2 Transition to new fibers

Various fibers of different performances and/or functions have been developed continuously. Although those fibers are referred to today as super-fibers, the name "super-fiber" was coined only during the 1980s, as illustrated in Fig. 1.2. Fibers before 1980 were either identified as general-purpose fibers or designated as speciality fibers. Carbon and aramid fibers were classified, for example, as speciality fibers.

High-value-added products of polyester fiber were first developed in Japan in the 1980s. Mono-filament of conventional polyester filament is 2–3 denier. Here 1 denier is defined as the weight of a filament of 9,000 metres in length, and was used to designate the coarseness of filament of raw silk or nylon. The coarseness of natural fibers such as cotton, ramie, silk and wool is expressed in terms of count, which is determined from the length of a 1 kg fiber. For example, one count is equivalent to a 1 km fiber of 1 kg weight. Cotton fiber of 240 counts is as fine as silk. Most men's shirts are made from blended fabrics of cotton and polyester, containing 35% cotton and 65% polyester,

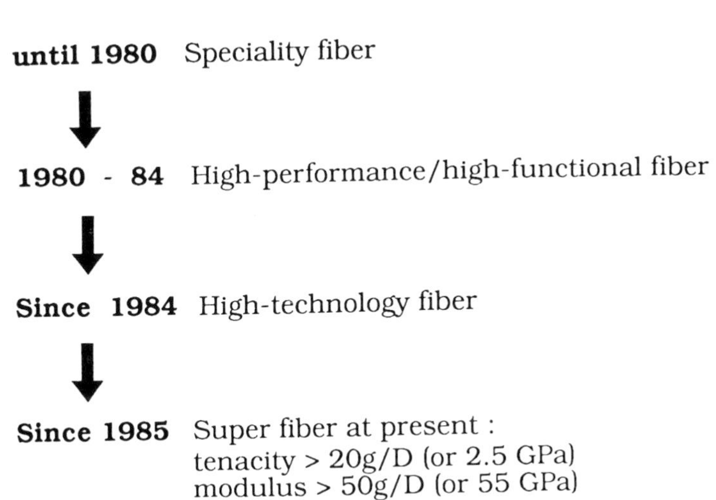

1.2 History of super-fibers.

although the fabrics of various blend ratios are commercially available according to the use and performance required. Polyester fiber, 1.5 denier, is normally used for blended fabrics, but additionally today, 0.5 denier fiber and 1.0 denier filament are available owing to further technical developments. Fibers of smaller denier or larger counts are technically more difficult to produce. Polyester filament became finer, reducing from 2–3 denier to the 1 denier produced now. Polyester fiber staple of blended fabrics for men's shirts or women's blouses came down from 1.5 to 0.5 denier. It is not always appreciated that it is technically difficult to spin fine polyester filament or staple. In fact, the spinning technique for forming synthetic fibers becomes more critical with shrinking denier, since the process requires three established technologies to spin homogeneously synthetic fibers as fine as human hair.

The three technologies that must be combined are:

1 Fine spinning.
2 Fine processing.
3 A highly reliable production technology to combine the above two technologies without producing any defect.

Engineers established these technologies as a result of consistent and intensive efforts over a considerable period. New products only come into existence when produced, sold and successfully evaluated by market forces. Synthetic fiber companies have accepted the challenge to spin finer and finer fibers. As the result of competition, many new products were put on the market, such as silk-like, bulky or good-to-touch textiles of nylon or polyester. These products are referred to as high-value-added or speciality products. Although the speciality fibers were deemed to be 0.5 denier, to distinguish them from conventional fibers of over 1.5 denier, consumers became confused and treated both high-performance fibers and high-value-added fibers of fine denier as speciality fibers.

Thus there was a move during 1980–84 to define the terms specifically in order to distinguish high-value-added fibers from high-performance/high-function fibers, specifying, on the one hand, polyester fibers of fine denier, and on the other, high-performance fibers such as carbon fibers and aramid fibers. High-performance/high-function fibers were later (in 1984) redeemed as high-tech fibers, as "high technology" became a term in common use. High-tech fibers now include new functional fibers such as biodegradable fibers, chemical absorbance fibers, fibers from biomaterials and activated carbon fibers, which are not of high modulus and high strength, but have other new performance advantages.

Therefore, high-tech fibers can now be redefined as fibers produced by high-technology, which are superior to those produced by conventional fiber technology, arising from the application of the newer developments in fiber science and technology. A survey on "high-tech fibers" and a book "high-technology fibers" appeared in the USA in the spring of 1985 (see Further reading at the end of the book). Since then, the term "high-tech fibers" has been generally accepted in Japan. This survey defines high-tech fibers as the fibers used in, and produced with, high technology. In this context, a super-fiber that is outstanding in certain physical properties such as high strength and/or high modulus can be defined as a high-tech fiber. Fibers with specialist chemical functions are not regarded as super-fibers. For example, high-performance fibers with biodegradable or chemical absorbance function are classified as "high-tech fibers" and not as "super-fibers".

This book reviews all new fibers, which include "super-fibers" and "high-tech fibers", classified according to the definitions given here, and additionally the new cellulose technology developed to process cellulose into speciality polymers. These moves have resulted as a direct consequence of cellulose being re-evaluated world-wide as a renewable resource following the oil crisis.

2 The super-fiber with new performance

2.1 Two streams of super-fiber

The Society of Fiber Science and Technology, Japan, organised ISF'85 (International Symposium of Fiber Science and Technology) at Hakone, Japan, in 1985 to celebrate its 40th anniversary. The main interest of the distinguished international gathering centred on the super-fibers, which were being developed competitively by many research institutes throughout the world. It is commonly thought that super-fibers emerged rather suddenly. Their development, however, was the result of merging fundamental scientific and technical knowledge which seemed at the time not related to fiber science. This was necessary to overcome a series of technical barriers. The history of science and technology teaches us that a new technology is frequently developed by a group of people obsessively concentrating on a particular problem. In this respect, a director of research and development needs to be an organiser, in addition to being a manager. Often, the future of an industrial development in high technology depends on a technical "conductor" who plans and leads the new technical development.

There are several influences in the development of super-fibers, among which two groups deserve special mention. One is the research group of the Du Pont Company in the USA, which developed the *para*-aramid fiber named "Kevlar". This material is seven times stronger than steel and, when it appeared, was the most sensational fiber since nylon. The second is the research group at DSM in The Netherlands, which developed polyethylene fiber of even higher tenacity than Kevlar. Du Pont started its research and development on heat-resistant polymers in the 1960s, in response to space development needs. A new process of liquid crystalline spinning was developed to spin Kevlar from rigid polymer by Kwolek's group in Du Pont. Gel-spinning was developed in the late 1970s by the DSM group headed by Smith (now at University of California, Santa Barbara, USA) and Lemstra (later Professor at Eindhoven Technical University) to spin extremely high molecular weight polyethylene. The molecular weight of polyethylene is of the order of 10^4 when used for plastic containers or bags. It is of the order of 10^6 for super-fibers, which are spun from gel-like solutions of high molecular weight polyethylene. Gel spinning was known in 1969, but its industrial application came in 1978 when DSM introduced polyethylene fiber of high strength. Later Allied Chemical Corp. in the USA developed polyethylene of high strength by another gel-spinning process. During ISF'85 at Hakone, scientists from DSM and Allied Chemical Corp. intensively discussed the mechanism of gel-spinning, and demonstrated the urgent need for a super-fiber of high strength and light weight.

Chemists have learnt a great deal from silkworms, and emulated them in spinning synthetic polymers. However, they have not yet been able to develop ambient temperature spinning, which would be more advantageous from an energy conservation viewpoint. For example, a silkworm swings its head in the shape of a figure 8 and spins silk fiber at room temperature. Although natural fibers such as silk, cotton and wool have been extensively investigated scientifically, the mechanism of their formation is not yet fully understood. However, the gap between the natural and the synthetic is now being continually reduced.

2.2 The quest for a strong fiber

Various functions are required from fibers. Each industry is developing its own fibers for specific functions. Increased high strength is always a goal of fiber scientists, who continually seek something stronger. A fiber should be strong not only with respect to its tensile strength, but also in its resistance against deformation (high modulus) and wear resistance. The development of higher-performance fiber for composite reinforcement is also urgently needed in space and ocean technologies. These are areas where there is intense competition between the European countries, the USA and Japan. Fiber strength can be defined either in terms of tensile strength, which controls mono-filament characteristics, or tear strength, which significantly influences film characteristics. However, it is deformation resistance, i.e. having a high Young's modulus, that is the most necessary property for composite enforcement. For example, a space shuttle can be rolling violently because of atmospheric pressures when plunging into the atmosphere, so it must be resistant to deformation. Young's modulus is defined as the ratio of the unit cross-sectional applied force (stress) to the strain in the force direction. It is, therefore, a measure of deformability, with larger values indicative of lower deformability.

High-technology composite materials are expected to serve as fundamental materials for supporting the technologies required for the 21st century, as demanded by the automobile/aircraft industries, and space technology. These will allow a significant extension to the range of human activity. As a by-product, the high-performance materials developed for aerospace technology have also opened up improvements in the fiber materials used in the leisure field, such as golf, tennis, skiing and sailing, particularly in Japan. Not only are these materials light and strong but they also provide the specialist performances needed by the particular application and/or environmental conditions encountered, as in space or on the ocean. Here the mechanical properties can be tailor-made by producing a composite, and as a result new advanced composite materials (ACMs) are being developed by many industries in Europe and USA. For example, carbon fiber possesses good tensile strength, but lacks impact strength, whereas high-tenacity aramid fiber is weak against compression. Thus the mechanical weaknesses of carbon fiber (low impact strength) and aramid fiber (poor compression resistance) can be compensated for by making a composite of both fibers in an epoxy resin matrix. This type of composite material is widely employed as an ACM of light weight for energy conservation in space and ocean developments. Composite materials are superior in physical properties and performance when compared with single materials, and are used in tonne quantities in aircraft. Composite materials could well account for about 60% of the primary and secondary structural materials used in aircraft, etc. in the 21st century.

These reinforcing fibers are assessed in terms of their modulus and tenacity. Super-fibers must have a modulus greater than 55 GPa and a tenacity of 2.5 GPa (see Fig. 2.1). GPa is an international unit of modulus or tenacity; 1 GPa corresponds to the strength equivalent to approximately 100 kg per 1 mm^2. Super-fibers such as high-tenacity polyethylene fiber, para-aramid fiber and PAN-based carbon fiber satisfy this condition which is required of a reinforcing ACM fiber. The tenacity of the super-fiber Kevlar is 25 g/denier, which is seven times stronger than that of steel of the same weight (3.5 g/denier). Kevlar even possesses a higher tenacity per cross-section than steel.

2.3 From "shish kebab" to "gel-spinning"

Shish kebab is a typical Turkish–Armenian grill dish of marinated lamb on skewers, where "shish" and "kebab" denote the skewer and meat, respectively. In 1966, Pennings (subsequently Professor at Groningen State University, The Netherlands), then at the DSM Company, observed a shish kebab-shaped crystal while stirring dilute solutions (of a few per cent) of high molecular

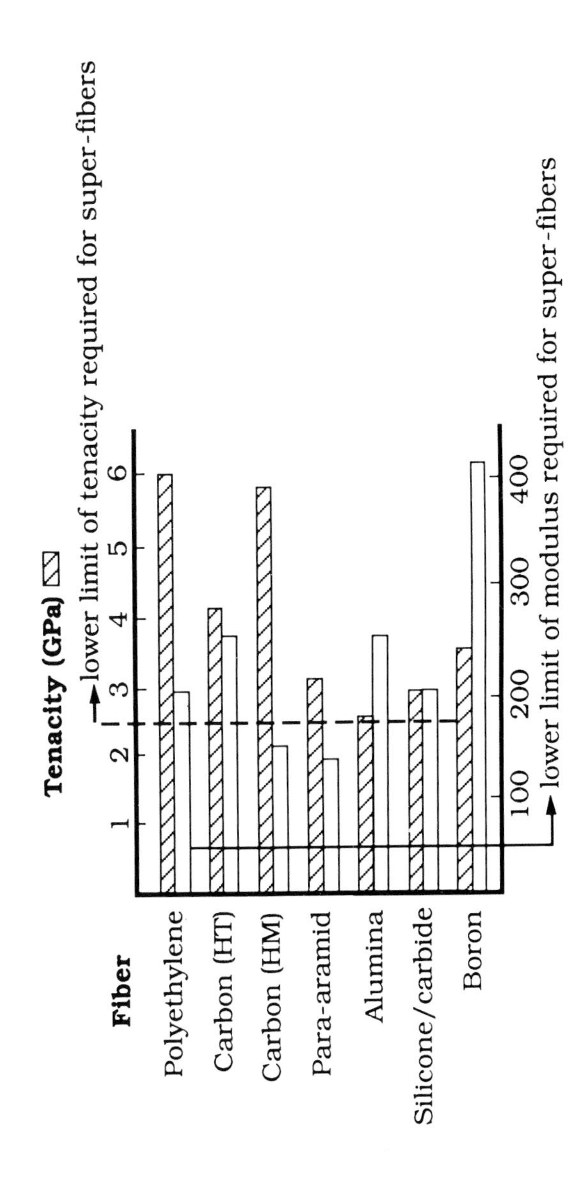

2.1 Physical properties of super-fibers.

weight polyethylene. He considered how he might make a high-tenacity fiber from this shish-part, and over many years worked energetically to separate the shish-part from the kebab-part, eventually succeeding.

Propane, used as a domestic fuel, is a gas with a low molecular weight of 44. This hydrocarbon compound becomes liquid and eventually solid like a wax candle when polymerised to increasing molecular weight. When its molecular weight becomes of the order of 10^4, the olefin compound, or plastic, is called polyethylene, which is rather hard and widely used for polybags and polyethylene bottles. There are now two types of polyethylene available for domestic use; one is the soft, branched, low-density polyethylene and the other is the hard, linear, high-density polyethylene. High-density polyethylene is produced using a polymerising catalyst at low pressure, and is used, for example, in making beer caskets. Super high molecular weight polyethylene can be classified as high-density polyethylene having a molecular weight of the order of 10^6. Its physical properties are very different from the ordinary 10^4 molecular weight product. Its chain length is almost a hundred times greater and its molecular weight two orders of magnitude greater. Polyethylene chains in this super-class product are coiled as in woollen yarn with greater entanglement. It becomes less fluid with increasing molecular weight, and it was first thought that this super high molecular weight polyethylene could not be spinned. However, in 1976, Pennings succeeded in continuously removing the shish-part, and Smith and Lemstra of DSM developed a practical method of gel-spinning by applying Pennings' technique. The method involves both spinning and drawing, in which a dilute solution (a few per cent) of polyethylene of high molecular weight is extruded into water to form a gel-like soft fiber, which is then heated and drawn out about 30 times in length. Ordinary polyethylene of molecular weight *ca.* 10^4 is drawn only up to 10 times in length and its Young's modulus is 1 or 2 GPa at most. However, the gel-like fiber of high molecular weight polyethylene drawn over 30 times can yield a Young's modulus of 90 GPa. This revolutionary method of "gel-spinning" attracted much attention from all over the world as a method for producing super-fibers.

Gel-spinning requires two steps. First, a dilute solution of polyethylene is prepared to reduce entanglement, and, secondly, a gel is formed from the solution to produce order in structure so that it does not entangle. The role of solvent is thus most important in gel-spinning. Drs Smith and Lemstra succeeded by using xylene or decalin as the solvent for this process.

Polyethylene is made up of regular crystalline regions and irregular amorphous regions when prepared by weak drawing, as shown in Fig. 2.2(a). When highly drawn upon heating, the crystalline region becomes oriented in

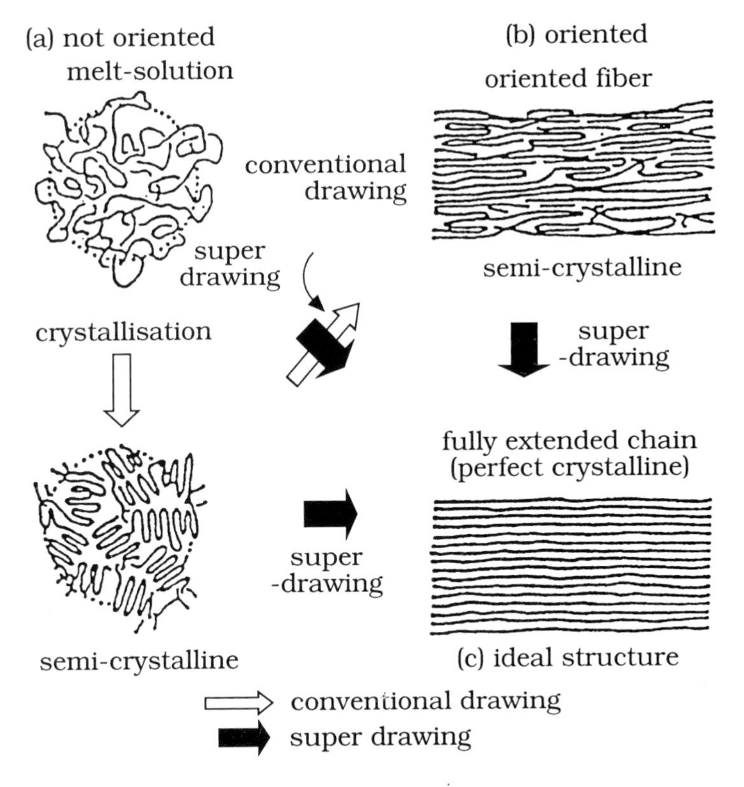

2.2 High-order structure of polyethylene.

one direction only (Fig. 2.2b). Further drawing increases the tenacity by approaching an ideal structure of extended chains, as shown in Fig. 2.2(c). Highly oriented polyethylene produced in this way can be used for packing tape, for example. An upper limit of the value of the modulus or tenacity of polyethylene can be calculated on the basis of an ideal chain structure of polyethylene molecules. Only 5% or 10% of this limit is reached in terms of the modulus or tenacity for ordinary polyethylene. Gel-spinning thus provided the first method capable of preparing polyethylene fiber with a tenacity and modulus that approached the theoretical maximum values. Although polyethylene molecules are flexible, they tend to crystallise when aligned in a parallel orientation. Once crystallised, the strong intermolecular interactions maintain the alignment of the polyethylene molecules. The problem, therefore, is how to align polyethylene chains in parallel alignment to make super-fiber of high modulus and high tenacity from conventional polyethylene.

This objective was achieved using the method of gel-spinning first patented world-wide by DSM. Although Toyobo Inc. (Japan) and Allied Chemical Corp. (USA) were pursuing their own research projects on gel-spinning simultaneously, they judged it impossible to escape from the basic patent applied for by DSM. They were forced, therefore, to enter into a technical association with DSM. Toyobo Inc. linked with DSM in March 1984, and set up a joint venture company Dyneema in The Netherlands in May 1986. In October of the same year, Toyobo Inc. completed a pilot plant at its Research Centre in Ohtsu, Shiga Prefecture, Japan, and started pilot-production of polyethylene super-fiber, eventually expanding the plant to commercial scale to produce 300 tonnes annually. DSM in The Netherlands increased production of polyethylene super-fibers to 1,000 tonnes per year in 1996 by constructing a new plant.

Allied Inc. (USA) was also investigating the production of polyethylene of high modulus and high tenacity independently. However, for the same reason, it was also forced to seek a licence for gel-spinning using the DSM process in 1984. Dr Prevorsek of Allied Inc. changed the solvent to paraffin oil, and developed high-tenacity polyethylene fiber Spectra 900 and later a higher performance fiber Spectra 1000. This material is used for a wide variety of industrial products such as helmets, suitcases and rope. Gel-spun Dyneema SK60 is a high-modulus and high-tenacity fiber of light weight, which is far superior in its impact strength to conventional fibers and exhibits the highest anti-fatigue breaking among the high-tenacity fibers. High-tenacity polyethylene is chemically stable and needs no special coating, but its melting point, varying from 145 to 155 °C, depends on the stressing conditions. It is also mechanically stable for a short period even at a temperature close to the melting point, and is used on its own as a fiber or as a composite material to reinforce other polymers. Considering its characteristics of light weight, high tenacity, anti-wearing, high ultraviolet light stability and electrical insulation, the ropes and cables made from high-tenacity polyethylene are expected to be used widely in future in the field of ocean industries and sports.

Dyneema SK71 is ten times stronger than steel in terms of the free breaking length, which is 428 km (see Fig. 2.3). Since the density of polyethylene is less than unity, its free breaking length is infinite in water. Fabrics made from this material possess high impact strength, high weather resistance and good hydrophobic properties, and as a result the material is widely employed for making bullet-proof clothing, protective wear, filters, sailing cloth and parachutes and for building materials. This high-tenacity polyethylene is also used as reinforcing material in composites for loudspeaker cones, archery bows and helmets, in order to exploit its high sonic propagation characteristics, vibration damping and high impact strength.

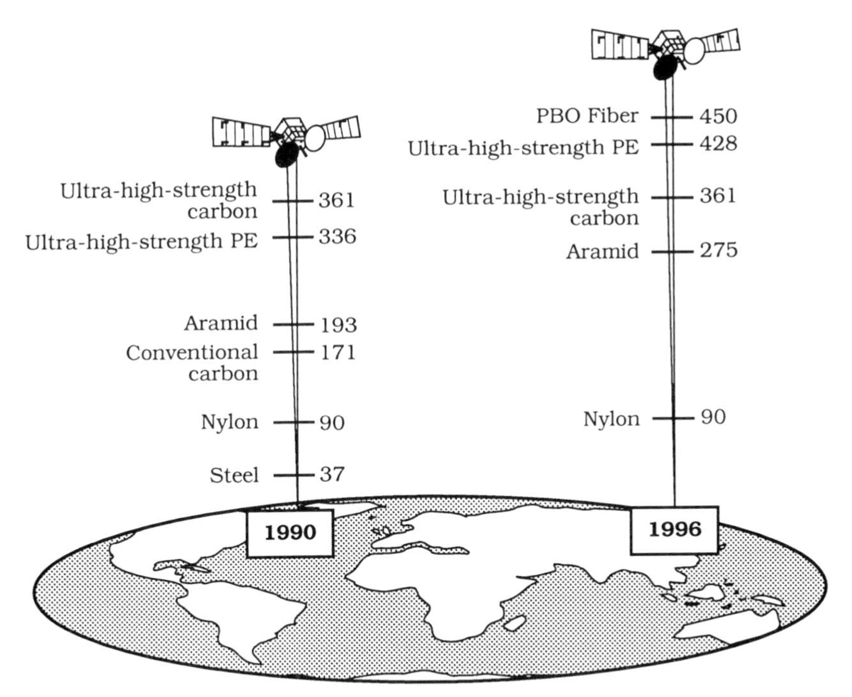

2.3 Free breaking length in air (km) of various fibers.

Mitsui Petrochemical Co. has developed a super high-tenacity polyethylene fiber Tekmilon, in collaboration with the Japan Research Institute for Polymers and Textiles, and has a pilot plant producing 5 tonnes per month in its Iwakuni plant. This polyethylene is used for the preparation of tennis racket gut, skis, playing clubs, bowstrings, FRP (fiber-reinforced plastics), etc.

Stimulated by the success of gel-spinning high molecular weight polyethylene, researchers in various countries have attempted the gel-spinning of other polymers, but without positive results. Allied Chemical Co., for example, developed gel-spun polyvinylalcohol (PVA) but not with the same success as polyethylene. Today Mitsui Chemical Co. of Japan, Hoechst of Germany and Hercules of the USA are commercially producing polyethylene of super high molecular weight, whose production amounts to 12,000–13,000 tonnes annually throughout the world. Nisseki Chemical Co. has developed a new method of mass production for high molecular weight polyethylene by combining its original know-how in polymerisation and catalysis techniques. The method, however, has not entered into full production since it does not proceed efficiently owing to the high viscosity of the polymer melt. The processing technology is still under development. Several other methods are being investigated currently to process high molecular weight polyethylene into fiber. Since the density of polyethylene (0.95) is considerably lower than that of Kevlar (1.45), high molecular weight polyethylene is more suitable for ACMs. The materials used in aerospace technology must be light as well as having high heat-resistance. Consequently, polyethylene cannot be used for this purpose.

2.4 The aramid fiber race in Europe, the USA and Japan

Commercial aramid fiber today now available is represented by Kevlar of Du Pont (USA) and Twaron of Akzo (The Netherlands). Approximately 2000 tonnes of Kevlar were imported and sold annually in Japan by Toray until domestic production of Kevlar started in 1988. Twaron is imported to Japan through the Japan Aramid Co., which is a joint corporation of Sumitomo Chemical Co. and Enka (a subsidiary of Akzo). Although a patent dispute between Du Pont and Akzo lasted for 11 years, it was eventually resolved. In Japan, Teijin Ltd. developed its own aramid fiber Technora, and started its commercial production in September 1987 at the Matsuyama plant, constructed in 1986.

2.4.1 Birth of liquid crystal-spun "Kevlar"

Space development projects started in the 1960s led to extensive joint research among industries, universities and governmental institutes in the USA to develop new materials. Among them, research and development leading to heat-resistant polymers by Du Pont was the most successful. For example, the heat-resistant polyimide resin Capton opened up a new field of application such as the construction of desk calculators, cameras and watches, in forms that are smaller, lighter, thinner and shorter. This is possible because the electric circuits can be printed on flexible polyamide film. Aromatic polyamide resins which have been developed by Du Pont include the zigzag-linked *meta*-type, called Nomex and the linear *para*-type named Kevlar.

The commercial production of Nomex was started in 1965, and this fiber is used for fire-fighting apparel, to allow fire resistance against flame, smoke and high radiation heat (see Fig. 2.4). This polyamide fiber is also used for protective garments worn at smelting furnaces or in oil refineries. The heat-resistant polymer has a molecular structure that is thermally stable, and stands up to long-term use at temperatures over 300 °C.

Developments always appear easier in hindsight. Nevertheless, their origination calls for enormous effort: it requires inspiration or a leap of the

2.4 Firefighter uniform of *meta*-type aramid fiber (Du Pont, Japan).

imagination to break through at a certain stage of development. Du Pont of USA developed in 1964 the *para*-type aramid fiber hailed as "...a new fiber of dreams stronger than steel..." which was then acclaimed as the most sensational fiber since nylon. The original development of this fiber originated with Dr Stefany Kwolek. Many researchers abandoned the task of dissolving the aramid resin in organic solvents, which were suggested by its molecular structure. Kwolek found, by accident, that a suitable solvent was one containing salt (sulphuric acid was eventually used in commercial production). She reported this finding to her supervisor, who did not believe her at first. When extruded from this dilute solution through a nozzle, there formed the epoch-making aramid fiber, which could hardly be cut by scissors. The term "epoch-making" is justified since this aramid fiber requires no drawing process, whereas conventional synthetic fibers such as nylon, polyester and acrylonitrile fibers must first undergo a drawing process after spinning through a nozzle. This drawing process is generally important in the synthetic fiber industry to furnish fiber in its finished form. Not only is the superficial shape controlled in this way, but the fundamental structure of the fiber will also be changed by drawing. During the drawing process, the molecular chains in polymer crystals will be aligned in the direction of drawing, and the Young's modulus and the tenacity of the fibers will increase.

Conventional synthetic fibers cannot attain sufficient strength without drawing, and would not find practical use. However, when aramid is spun from this concentrated sulphuric acid solution, the fiber shows high tenacity/high Young's modulus without the need for drawing and additionally exhibits high heat resistance.

Kevlar is spun from sulphuric acid solution where extremely rigid Kevlar molecules form a lyotropic mesophase. When the liquid crystals are extruded through a nozzle, Kevlar chains orient in the direction of the fiber axis and form fiber in the coagulating bath. Drawing is, therefore, not required to increase the tenacity by chain orientation in the direction of the fiber axis. This form of spinning is referred to as "liquid crystal spinning". The American Chemical Society Prize for Creative Invention in 1980 was awarded to Stefany Kwolek for her invention of this new process. Kevlar produced by this new technology of "liquid crystal spinning" provided an entirely new super-fiber which exhibits extremely high tenacity/high modulus as well as high heat-resistance. The impact of Kevlar proved so dramatic that researchers throughout the world rushed to synthesise rigid polymers and their subsequent liquid crystal spinning. Du Pont alone invested US$300 million over 10 years to get the aramid fiber Kevlar on to the market. The term "aramid" was adopted for the allaromatic polyamides first by the US Trade Committee in 1974, and subsequently approved by the ISO (International Organization for Standardization) to distinguish the group from the aliphatic polyamides (nylon 6). Although the commercial success of aramid fiber cannot be attributed to one particular person, Jeorge Lanzel, Managing Director responsible for the textile/fiber R&D of Du Pont, must surely take the main credit for this success. One might comment that Du Pont was well equipped with talent, funds and facilities, but these assets are only part of the story. What is important is a readiness to cultivate the small seed of an idea into a large tree that bears fruit. In this respect, the role of a technical coordinator, such as the head of the R&D section, is vital to stimulate and control talented researchers without suppressing their interest. No success can be expected in industry without the harmony of stimulation and control. R&D in Japan is often said to be no more than yielding a small-scale technical improvement. The Japanese are essentially a conservative farming race and a group consensus weighs more in their decision making than individual opinion. Once a potentially creative young Japanese of strong individuality enters the well-organised industrial society, he or she soon becomes accustomed to the environment and aims towards being a manager rather than using his or her creative talent. This might well be a consequence of the uniform education system in Japan, whereas in Europe and the USA individual creativity seems to be given greater scope in the educational system. The Europeans are a hunting race, and prefer to be different from others, with respect to originality. Since industrial internationalisation is inevitable, creative talent should be valued more in Japanese industries.

Du Pont produces Kevlar at its plant on the outskirts of Richmond in Virginia, and production capacity was increased to 21,000 tonnes/year in 1996. It also started Kevlar production at a plant in Maydown, Northern Ireland, in 1988. Production in 1996 was 5,000 tonnes per year. Toray–Du Pont, a joint company of Toray and Du Pont, has been importing Kevlar to Japan since January 1985, and it is sold through Toray Co. A newly joint company of Toray–Du Pont Kevlar was established in February 1989. A building was completed in 1991 in Tokai, Aichi Prefecture, Japan, to house a new production facility for Kevlar (2,500 tonnes/year). There are plans now to extend the application of Kevlar to produce rope for special uses, cables for optical fiber, safety helmets for sports and motorbikes and other wear-resistant materials. Approximately half of the Kevlar produced in Japan is used for cords of radial tyres for passenger cars, and the rest for industrial and sports/leisure materials such as protective working clothes and gloves, and tennis, golf and sailing equipment. Kevlar's major use throughout the world in 1984 was for composite materials in aerospace technology (51%) and tyres (21%). Since the consumption of ACMs has increased 30% annually over the past 10 years, Du Pont is actively expanding its capacity for Kevlar production. Du Pont also expects the ACM markets for aerospace technology, car industries, general industrial parts, construction materials, ceramics and electronics to increase dramatically. While *para*-type aramid fiber is mostly applied to heavy industrial uses, its physical properties are attractive for the clothing industry also. It is used for special clothing such as bullet-proof jackets, safety jackets and protective gloves. Aramid fiber is five times stronger in terms of the tensile strength and one-fifth lighter in terms of the density than steel, and exhibits high wear-resistance. Goldwin, a sportswear maker, uses aramid fiber to make trousers for mountaineering and anoraks of blended fabrics with wool or cotton. Kevlar–wool blended fabrics with 10 wt.% Kevlar are used for autumn/winter trousers, and Kevlar–cotton blended fabrics with 40 wt.% Kevlar (Kevlar filling yarns and cotton warp ends) for spring/summer trousers and anoraks. Kevlar is being expanded now into the aerospace, car, and optical communication industries, and for producing safety/protective clothing, wear-resistant materials, and ocean technology, construction materials, etc to exploit its special physical characteristics. After the big Kobe earthquake of 1995, concrete structures such as highways, railroads, subways and buildings needed to be quickly reinforced. Aramid fiber was one of the materials used because of its high tenacity and modulus, ease of working and its non-conductivity. Kevlar sheets were applied to reinforce highway columns, chimneys, lighthouses, etc. It will shortly also be applied to railway columns.

2.4.2 A new product: "Kevlar 149"

Du Pont has succeeded in developing a new *para*-type aramid fiber, Kevlar 149. Original Kevlar material is polymerised from *para*-phenylenediamine and terephthalic acid chloride using a catalyst. Wet/dry spinning from sulphuric acid solution yields highly oriented Kevlar 29 without drawing. Kevlar 149 is produced by heat processing of Kevlar 29. Kevlar 149 exhibits approximately 40% greater tensile modulus and 50% less moisture absorption than other high-modulus aramid fibers. It is thus able to replace steelwire cords in the rope/cable industry because of its high tensile strength and good creep characteristics. Kevlar 149 is a new generation that exhibits special characteristics as an advanced composite material with carbon fiber in applications that require high modulus and light weight.

2.4.3 Twaron

Akzo Co. of The Netherlands started research on the *para*-type aramid fiber in 1969, and as a result developed Twaron. Enka Co., a subsidiary of Akzo, built a pilot plant in 1976 to promote production and applications of Twaron. A Twaron plant with a production capacity of 5,000 tonnes/year (expandable to 10,000 tonnes/year) was completed in Emen, The Netherlands, at the end of 1985, and commercial production was started a year later. This production process for Twaron is similar to that of Du Pont, but Enka uses a less expensive solvent. The physical properties and special characteristics of Twaron are also similar to those of Kevlar, apart from one: it can yield an activated aramid which is preprocessed to increase the adhesion towards rubber. Akzo has developed strongly in Japan; the Japan Aramid Co. is a joint company formed by Enka and Sumitomo Chem. Co., which has imported and sold Twaron since January 1987.

2.4.4 Technora

The co-polymer type *para*-aramid fiber Technora was developed by Teijin Ltd in 1974. The production plant in Matsuyama with a capacity of 700 tonnes per year was completed in September 1994. Production capacity will be increased to 1,500 tonnes per year by 2,000 (see Fig. 2.5).

Aromatic polyamides (aramid fibers) are prepared basically from

2.5 Technora of 11 mm ϕ lifts 8 tonne tetrapot (Teijin Ltd).

terephthalic acid chloride and another component amine that has a great influence on the physical properties and cost of the final product. Du Pont employs petroleum-derived terephthalic acid chloride and *para*-phenylenediamine to produce Kevlar, whereas Teijin introduced diamine as a third component. Technora is wet-spun and drawn instead of liquid crystal spinning. Technora includes ether bonds in its molecular structure, and has higher tensile strength, better resistance against fatigue and abrasion, higher heat resistance for long period and better chemical resistance than Kevlar and Twaron. Technora is available in various types and forms including dope dyed black and gray colors, stretch broken yarn using specialised fine denier of 0.75 denier, and monofilament yarn in 8–100 denier. These properties are described in Table 2.1 together with the appropriate application.

2.5 Polyacetal fiber

In 1959, Du Pont commercialised polyoxymethylene (POM) as polyacetal resins under the trade name Derlin, which is now used for gears and bearings of videocassette recorders, the reels of magnetic tapes and clock gears. Du Pont introduced the description *"A challenge to metal"* for this new

Table 2.1. Various characteristics of Technora

	Field	Possible applications	Applicable type
Rubber reinforcement	Tyre	Car, truck and racing tyres	T-200
	Belt	Belt for continuous variable transmission synchronous belt, V-belt, conveyor belt	T-202 T-203
	Hose	High-pressure hose, steam hose, radiator hose, heater hose	
Industrial material	Rope, cable	Mooring line for oil rig, electrical engineering rope, sling rope, antenna stay, yacht rope, sports net rope, rope grip	T-220 T-221 T-230
	Cord, braid	Sewing thread, fishing thread, tennis gut, archery bow, heater wire cord, earphone cord, paraglider cord, optical fiber tension member	T-240
	Woven belt		T-240
	Narrow fabric	Heat-resistant belt, sling belt, safety belt, tape	T-360 T-370
	Industrial fabric	Membrane structure, tent, dryer-canvas, explosion protective sheet, bag material, yacht sail, parachute	T-240
	Net	Knotless fishing net, safety net	T-240
	Filter	Heat-resistant filter, acid-proof filter	T-240
	Non woven fabric	Heat-resistant belt	T-330
	Civil engineering	Banking reinforcing grid net, anchor-bag, asphalt-reinforcing fabric, seabed subsidence protection fabric	T-240 T-360
Protective material	Anti-ballistic	Armour vest, helmet shell, armour plate	T-240
	Cut resistant clothing	Safety gloves, apron, working wear, shoes, sportswear	T-240 T-360
	Anti-melting clothing	Spatter-resistant clothing, rider suit	T-370 T-330
Asbestos replacement	Friction mate	Brake pad, clutch facing	T-240
	Gasket	Engine gasket	T-320
	Packing	Gland packing	T-340
Cement reinforcement	Building materials	Curtain wall, floor material, ceiling material	T-320
	Civil engineering material	Pipe, prestressed tendon	T-321 T-240
Plastic reinforcement	FRP (thermosetting type)	Aircraft parts, sports goods, industrial equipment parts, pressure vessel, printed circuit board	T-320
	FRTP (thermoplastic type)	Lubricant parts material, business machine parts, electronic equipment parts, casing	T-322

plastic material. Asahi Chemical Industry applied microwave oven technology, and developed a new technique (dielectric heating/drawing method) to produce a wire-like polyacetal fiber "Tenac SD". This method was originally invented for reinforcing materials for optical fibers at the Ibaragi Institute for Electronics and Telecommunication Research of NTT

(Nippon Telegraph and Telephone Co.). Although the drawing rate was initially too low for industrial purposes, Asahi Chemical Industries succeeded in developing a new technique of super-drawing (see below) for polyacetal filament. In this way, fine as well as thick filaments can be produced. Since crystals are aligned regularly in thick filaments, they are rigid like steel wire, and hence are called "wire filaments". Super-drawn polyoxymethylene filament is now also prepared by continuous super-drawing under high pressure.

When stretched, commercial polyethylene film undergoes necking suddenly and then stretches further until it breaks. The further drawing after necking is called "super-drawing". It is difficult to draw filament ten times its length after necking. The optimum conditions for polyacetal were found to be continuous super-drawing under high pressure.

POM is placed in contact with pressurised liquid and drawn while passing through its container. Transparent polyacetal filament (0.5–2 mm in diameter) prepared in this way has a fine structure and high modulus/high tenacity. No filament made from thermoplastics, including polyacetal, has a fully extended chain structure, but the polyacetal filament prepared in this way is composed of whisker-like units. The "whisker" denotes a short, staple-like single crystal. The polyacetal filament has a fine structure of well-oriented chains, and its mechanical strength is higher than that of steel wire. Its heat-resistance, chemical stability and weather-resistance are excellent. Polyacetal offers a material, therefore, to replace steel wires and ropes. POM filament is strong enough to withstand shark's teeth, when employed in longline fishing. Since POM is stable against alkali, its filament can be used also as a concrete-reinforcing material. POM is not attacked by moths, nor does it get mouldy. Seaweeds, shells or ocean microorganisms will not grow on POM as on conventional fishing nets, which have, therefore, to be coated with organic zinc, etc. No pollution problems arise from the use of POM nets. POM filament exhibits excellent creep characteristics and does not absorb water. When struck, it produces a metallic sound. Thus, highly elastic POM filament is ideal for tennis gut, which must provide a good feel as well as a good sound when hit by the ball.

2.6 Strong Vinylon RM

Motivated by the gel-spinning process developed by DSM, Kuraray Co. produced high-modulus and high-tenacity Vinylon RM by gel-spinning polyvinylalcohol containing boric acid, after spinning and drying/drawing at high temperatures. Conventional Vinylon is used for non-woven clothes for industrial use and for fishing nets. Vinylon RM has better mechanical properties; its surface is rough so that the adhesion to hard materials, such as cement, is good. It is also hydrophilic and alkali-resistant, and disperses well because of its specially processed surface. Vinylon RM finds application in areas such as Vinylon-reinforced slate and Vinylon-reinforced mortar cement.

PAN and PVA fibers in general use have the potential to yield high-modulus and high-tenacity fiber. PAN can achieve a high degree of polymerisation, but not PVA. Allied Inc. of the USA has developed high-modulus and high-tenacity PVA of super high molecular weight by gel-spinning. Dr Prevorsek of Allied Inc, first described the gel-spinning of PVA and the difference between the gel-spinning mechanism of polyethylene and PVA at a Post Symposium to ISF'85, held in Hakone. He also reported that the mechanical properties of PVA fiber with fully extended chains were similar to those of aramid fibers. Now many research institutes throughout the world are actively pursuing the development of high-modulus and high-tenacity fiber from polycondensates such as nylon or polyester.

2.7 New liquid crystalline polymers: engineering plastics

There is currently much interest in liquid crystalline polymers as possibly polymeric materials, and as a result new products utilising liquid crystalline polymers have appeared on the market from various companies (see Table 2.2). These liquid crystalline polymers are expected to replace metals, especially those used for automobile, electronics and electrical parts because of their characteristics of high modulus/high tenacity, high heat-resistance, good chemical stability, high processing precision and light weight. There is promise that they will be the engineering plastics for the next generation.

Liquid crystalline polymers are classified as either lyotropic or thermotropic liquid crystals. The former yield liquid crystals by adding solvent at a certain temperature and concentration, whereas the latter form utilise thermal changes like cooling/heating. In the nomenclature "lyo" and "thermo" denote "dissolve" and "heat", respectively, and the "tropic" corresponds to the "property". Aromatic polyamides and aramid, represented by Kevlar, belong to the lyotropic liquid crystalline polymer type. Since these polymers have a high melting point, which is close to their decomposition temperatures, they are normally spun from a solution in a special solvent such as sulphuric acid. The thermodynamic liquid crystalline polymers include aromatic polyesters. Unlike lyotropic liquid crystalline polymers, thermotropic liquid crystalline polymers can be processed in a variety of ways, for example, injection moulding, extrusion moulding and melt

Table 2.2. Names of producers and products of liquid crystalline polymers

July	1979	Sumitomo Chemical Co	Sumicasuper
December	1984	Dartcore, now Amoco	Xydar
October	1985	Mitsubishi Engineering Plastics Corp.	Novacuurate
October	1985	Unitika	Rodrun
December	1986	Polyplastic Co. and Celanese Corp. (now Hoechst Celanese)	Vectra*
May	1986	Kuraray–Celanese	Vectran
	1987	Japan Petroleum Chemical Co.	Xydar
March	1996	Toyobo Co.	Zylon

*Denotes the products include fiber and plastics. Fiber-only products are available for Vectran.

spinning. In December 1984, Dartcore Co., USA, developed the first commercial thermotropic crystalline polymer, Xydar, suitable for melt moulding. The spinning of aromatic polyester is relatively simple, whereas fully aromatic polyamide (aramid), represented by Kevlar, must undergo several complicated steps from the original polymer to the final fiber product. Consequently, thermotropic liquid crystalline polymers have attracted much interest, purely from an economic standpoint. A number of liquid crystalline polymers of the polyester type are now commercially available, as summarised in Fig. 2.6. Among them, Xydar of Dartcore Co. and Ekonal of Sumitomo Chem. Co. are derived from Ekkcel of Carborandom Co., USA, whereas Novacuurate of Mitsubishi Chemical Corporation and Rodrun of Unitika Ltd, are based on the X7G polymer of Eastman Kodak Co., Polyplastic Co.; Kuraray employs Vectra from Celanese Corp. for its products.

These fully aromatic polyesters are classified into three types, based on their physical properties. The first type, including Xydar and Ekonol, are characterised by their high heat-resistance; the second type, Vectran, offers easier processing without sacrificing heat-resistance and mechanical strength. Chain orientation (anisotropy) is much reduced in the third type, comprising Novacuurate and Rodrun. Mitsubishi Chemical Corporation and Unitika developed the coating material for optical fibers from aromatic polyester in cooperation with NTT. They expect their product to be used for electric/electronic parts (connections, bearings, etc.), film and textiles. For example, Unitika has developed LC-5000 and is investigating how the spinning of this material can open up new markets. Liquid crystalline fully aromatic polyester is thus an engineering plastic for the 21st century, which can be used either as resins or fibers.

Jell-O is a food product whose texture is different according to the ingredients mixed within the gel substrate. If fully aromatic polyester resin is compared with Jell-O, the gel corresponds to the *para*-hydroxybenzoic acid

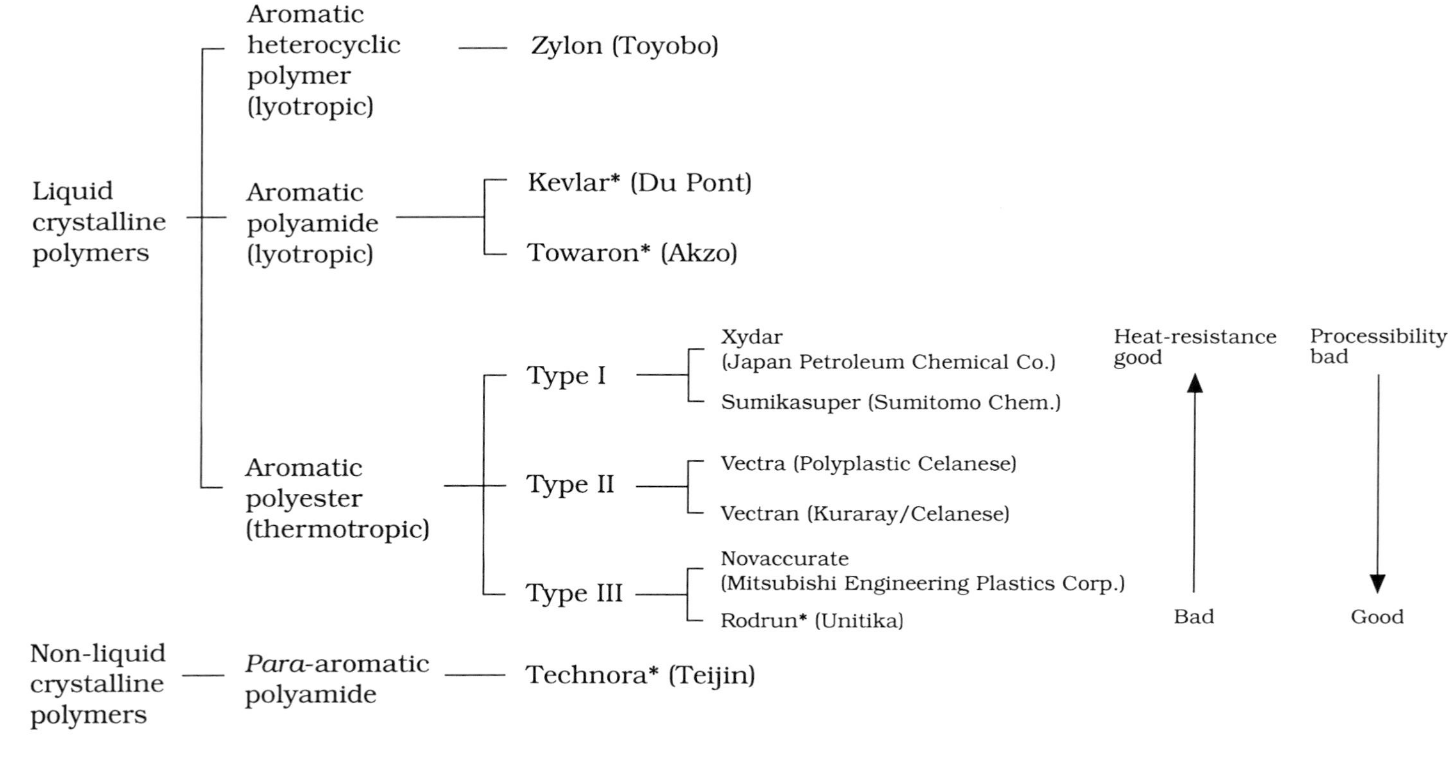

2.6 Types of liquid crystalline polymer. * denotes the products that include fibers and plastics. Fiber-only products are available for Vectran.

ingredient and the other ingredients to the other polyester components, which determine the physical properties and the cost of fully aromatic polyester resins. Vectran of Kuraray and Celanese employs fine hydroxynaphthoic acid as another ingredient within *para*-hydroxybenzoic acid "gelatine" component to produce fully aromatic polyester "Jell-O". Coarse dihydroxy-diphenyl and a small amount of the third component terephthalic acid are mixed within *para*-hydroxybenzoic acid in the products Ekonol of Sumitomo Chem. Co. and Nippon Exran Co. There are various ingredients used to make thermotropic fully aromatic polyester "Jell-O", to allow them to exhibit a different texture (in this instance thermal response). The thermotropic liquid crystalline polymer is quite different in nature from nematic liquid crystals (although both are classified as "thermotropic") of low molecular weight which change their degree of transparency according to applied voltage. These are employed in digital display panels for desk calculators and watches. Here "nema" means "threadlike", since a threadlike pattern is observed from this type of liquid crystal when viewed through a polarising microscope. Fully aromatic polyester is thermotropic, and assumes a liquid crystalline structure at high temperatures. The melt viscosity of fully aromatic polyester is comparatively low owing to its liquid crystalline structure, and its spinning is relatively easy.

Although both fully aromatic polyester and aramid are classified as liquid crystalline polymers, the physical characteristics of the respective polymers are different and so are the spinning processes applied to them. Kevlar must be "wet-spun" in the coagulation bath with a low spinning rate to recover its solvent sulphuric acid, whereas fully aromatic polyester can be "melt-spun" using a spinning rate of 1,000–1,500 m/min which is generally used for conventional synthetic fibers. It could be that because of the superior economic and technical aspect of this different spinning process, fully aromatic polyesters are getting more popular, and are replacing their predecessor aramid super-fiber. However, this simple comparison may not be truly valid when the whole process from the original polymer to the product fiber is considered. Fully aromatic polyester needs heat treatment of 300 °C or more after the spinning process, and the mechanical properties improve, usually because of further polymerisation due to this heat treatment. The physical properties of aramid and fully aromatic polyester fibers are summarised in Table 2.3. Fully aromatic polyester fiber is comparable to aramid fiber in terms of the modulus and tenacity as shown in Table 2.3. For example, Ekonol exhibits higher modulus and tenacity than Kevlar. No commercial sample of fully aromatic polyester fiber is available yet on the market.

Table 2.3. Physical properties of super-fibers

	Fully aromatic polyester		Fully aromatic polyamide (aramid)			Aromatic heterocyclic polymers (polybenzazole)
	Econol (Sumitomo Chem.)	Vectran (Kuraray)	Kevlar 29 (Du Pont)	Kevlar 49 (Du Pont)	Technora (Teijin)	Zylon (Toyobo)
Density (g/cm^3)	1.40	1.41	1.44	1.45	1.39	1.56
Tensile strength (g/denier)	30.8	27	22	23	25	42
Tensile modulus (g/denier)	1.080	600	525	800	570	2000
Breaking elongation (%)	2.9	3.8	4.0	2.7	4.4	2.5
Water absorption (%)	0	0.01	7.0	2.0	3.0	6.0

2.8 Vectran: a fully aromatic polyester fiber

Kuraray developed the high-tenacity fully aromatic polyester fiber Vectran (see Fig. 2.7) in cooperation with Celanese, and then started its consumer research. Kuraray already produces polyester fiber and Vinylon, but it was thought that these fibers would not compete in future from the point of view of tenacity and Young's modulus. Values in excess of 20 g/denier and 500 g/denier will be required for tenacity and Young's modulus, respectively, whereas these values for polyester fiber and Vinylon are 12 g/denier and 250 g/denier. Kuraray selected Vectran as its high-performance fiber in order to utilise its experience with polyester fiber spinning. Since the consumer research results turned out to be promising, Kuraray decided to start full-scale production (400 tonnes/year) of Vectran fiber and its marketing from February 1990. In USA and Europe, Hoechst–Celanese, in cooperation with German Hoechst (in the case of Europe) has conducted consumer research leading to full-scale production. The valuable property of Vectran, according to Kuraray, is that it absorbs no water and undergoes no physical change in the wet state. For example, the bonding strength between layers of Vectran and the matrix resin will not deteriorate under the influence of moisture when used as reinforcing fiber materials for FRP. This characteristic also prevents any size change of the timing-belt due to the repeated moisture absorption/desiccation when in use.

A second characteristic is its good shock-absorbency, which is valuable when using ACM for skis, PC boards and loudspeaker cones. The shock-absorbency of super-fibers is best for Vectran, followed by aramid fiber and

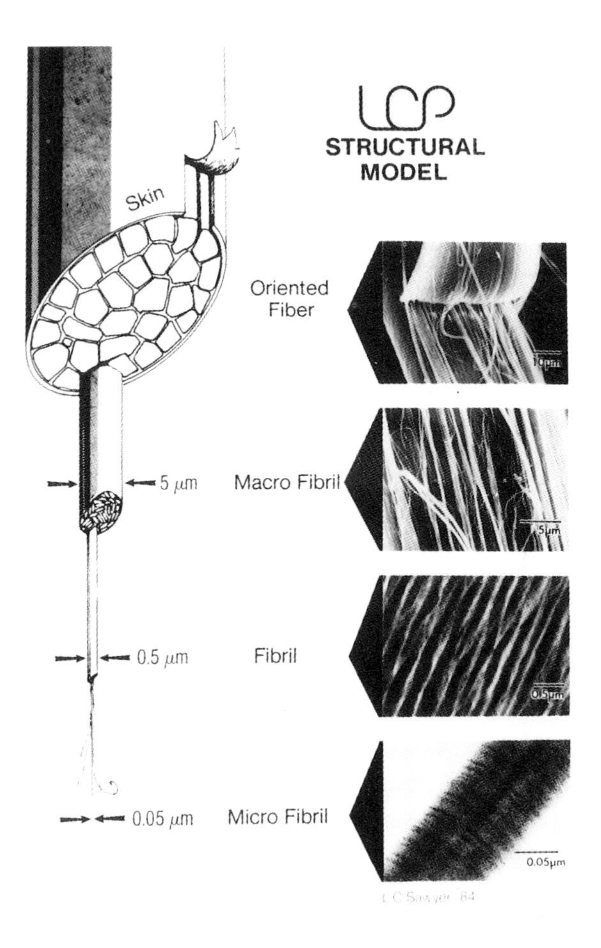

2.7 High-tenacity polyallylate fiber Vectran (Sawyer, L. C. and M. Jaffe (Journal Mater. Sci., **21** (6), 1897 (1986)).

then carbon fiber. High durability represented by high wear resistance and high-heatage resistance, and secondly high chemical stability (especially against acids) are other characteristics of fully aromatic polyester which guarantee a long life for Vectran products such as ropes, gloves, protective clothes and acid-resistant filters. Vectran costs US$45/kg, and only one type is commercially available at present. However, Kuraray has developed a new type Vectran of higher modulus to match Kevlar 49. Vectran products are available in the form of multi-filament, chipped strand, pulp, spun-like yarn and mono-filament. Vectran is being applied in the field of industrial materials such as tyre-cords, rubber materials (belts, etch ropes, tension members for optical fiber, friction materials, fenders, composites, etc.). Up until now aramid fibers include Kevlar of Du Pont, Twaron of Akzo and Technora of

Teijin have monopolised the super-fiber market. However, fully aromatic polyester fiber may well break through this monopoly. Hot competition will start in this market with the outcome dependent on the total research and development power, cost performance and marketing ability of each company.

2.9 Developing polyallylate fiber

The development of highly crystalline polymer materials is one of the main objectives of the Research and Development Project "Basic Technology for Future Industries" (Jisedai), organised by the Ministry of International Trade and Industry, Japan. The largest 11 textile/chemical industries, including Toray and Asahi Chemical Industries, in May 1984, formed the Research Association for Basic Polymer Technology. They have now developed high-modulus polyallylate fiber which belongs to the fully aromatic polyester category. This project aims to develop polymer materials of high heat resistance, high modulus and high strength to replace metals. Polyallylate assumes a thermotropic liquid crystalline structure around 300 °C. Its fiber of 0.12 mm in diameter exhibits a high modulus of 95 GPA, which is higher than that of Kevlar of the same diameter. It is not yet decided which company will produce polyallylate fiber on a commercial basis. Several of the companies in the consortium are willing to industrialise it when the social environment is suitable.

2.10 The ACM industry in the USA

Toray and Toho Rayon now compete with each other for the premier material supplier position of PAN-based carbon fiber, which can be regarded as the senior super-fiber in the aramid category. The history of carbon fibers began in 1959, with the pyrolysis of rayon in USA, which achieved a brief moment of glory in the UK, but was commercially developed in Japan as PAN-based carbon fiber. During this period, rayon-based carbon fiber in the USA failed commercially, whereas PAN-based carbon fiber in Japan made great strides. Now pitch-based carbon fiber is showing commercial expansion after a long induction period. The present world production capacity of PAN-based carbon fibers amounts to 8,930 tonnes/year and Japan produced 3,840 tonnes/year of this in 1990. It is evident that only half a dozen or so PAN-based carbon fiber industries in the world will be able to survive and flourish in the 21st century.

The ACM industry played a key role in the competition for space developments in the USA in the 1960s, and has grown rapidly since. However, there has been a drastic rearrangement of ACM industries from 1983 to 1985 to ensure their survival. Since ACM requires a long-term as well as a wide range development of technologies, including materials, processing and applications, many companies have retreated from the field and sold their ACM plants to big chemical industries in Europe and the USA at a high premium. For example, Du Pont, USA bought the carbon fiber section of Exxon Corp. in August 1984; BASF, Germany, acquired the Seryrion carbon fiber plant of Celanese (built in cooperation with Toho Rayon) in April 1985; ICI, UK, bought a subsidiary of Beatrix Co.; and Amoco, USA, took over the carbon fiber section of UCC in 1985 to develop its ACM activities in the future. These trends show that the ACM industry is not an easy business, despite the fact that the product seems to have a bright future. To be flourishing in the 21st century, the ACM industry still needs a large amount of working capital, accumulated technical know-how and experienced staff for research and development, which only the big chemical industries can afford.

PAN-based carbon fiber is mainly applied in aerospace technology in the USA, where the defence industry was the driving force in the expansion. CFRP (carbon fiber reinforced plastic) amounts to 28% of the total weight of the structural materials of the Harrier fighter. It comprises 30% of the B1 or B2 bombers which may increase to 60% if CFRP is employed for the primary structural materials of the bombers also. The economic benefits of reducing the weight by 1 kg in the space/aircraft industry are dramatic, as shown in Table 2.4. M. J. Sulkind of the US Air Force reported at the third US–Japan Meeting on Composite Materials held in Tokyo in July 1986 that more than 50% of the structural materials in the bodies of US fighters of the next generation (ATF) and of supersonic aeroplanes would be composites. Of the civil Boeing 767 aeroplane frame, 5–10% is constructed of composite materials, an epoxy matrix reinforced by carbon fiber and/or aramid fiber (see Fig. 2.8). More carbon fiber reinforced composite materials will be employed in the new Boeing model B747-400, which was first commercially available late in 1988. These materials will also be applied to the primary structural frame, such as main wings and bodies, as well as the secondary structural frame such as ailerons and doors of the Boeing 767 commissioned in 1992.

Composite carbon fiber materials and composite glass-fiber materials have been improved remarkably in tenacity, anti-impact, anti-abrasion, and

Table 2.4 Economic benefits (US$) of reducing the weight by 1 kg

Rockets/satellite	200,000
Missiles	1,500
Helicopters	350–1,500
Aircrafts	70–350
Ships – Industrial Materials	40
Cars	less than 4

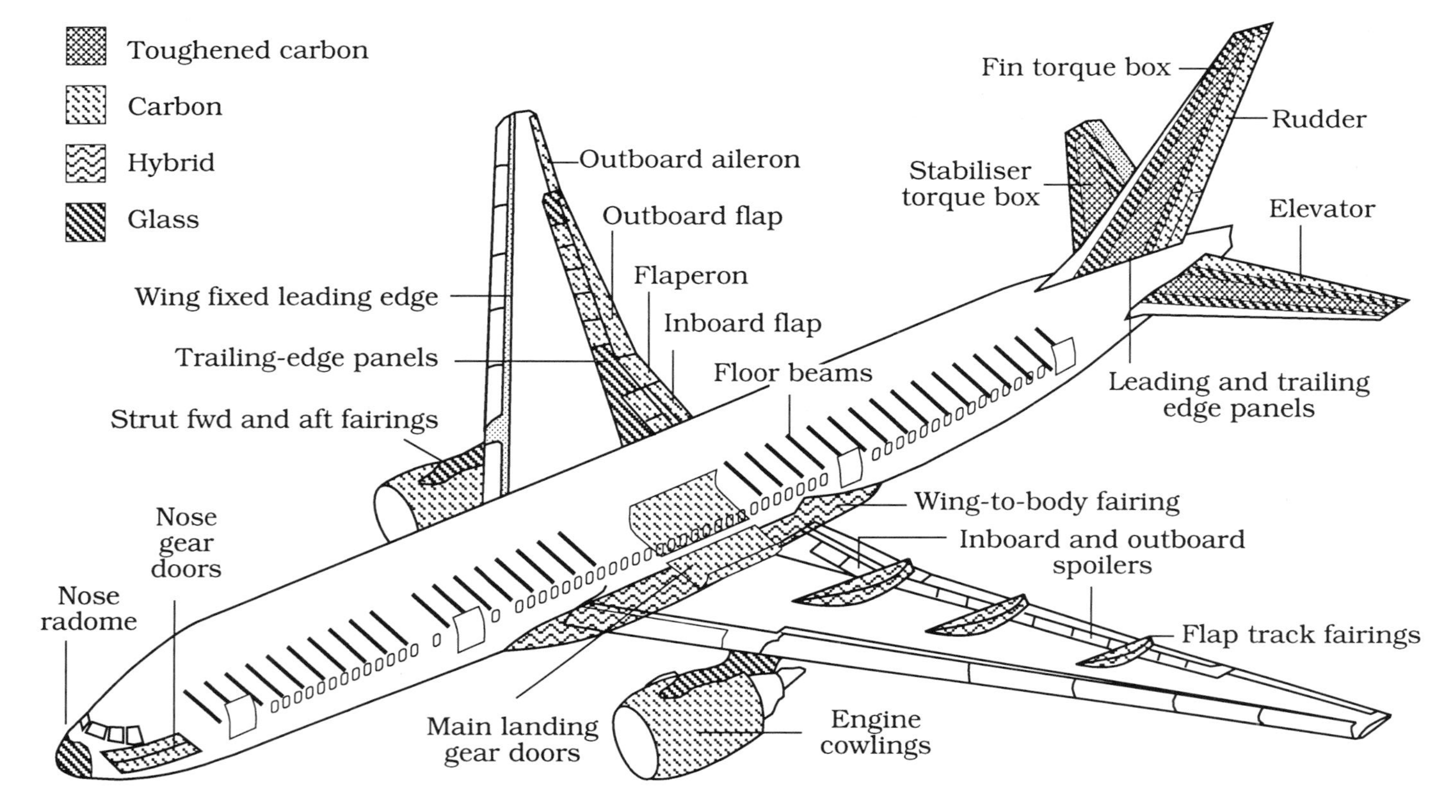

2.8 ACMs to enable weight reduction in aeroplanes (Boeing 777).

anti-corrosion properties. The weight of carbon fibers used in the former model of the aeroplane (Boeing 767) was 3% of the total weight. However, it increased to 15% in the new model (Boeing 777), expanding the fibers' use for the cabin floor, and horizontal and vertical tailpieces. The use of carbon fibers in aeroplanes contributes to a reduction of fuel consumption, owing to the decrease in weight of the aeroplane. The amount of CFRP used in the two Boeings is shown in Table 2.5. Here the performance of the matrix resin is more critical than that of the reinforcing fibers. A new thermoplastic resin other than the epoxy type must be developed urgently to improve the impact strength, heat-resistance, flame-resistance and solvent-resistance. Since there is no resin available that produces no poisonous gas while it burns, highly flame-resistant resin is also needed to improve safety. Now, polyether etherketone (PEEK) and polyphenylene sulphide (PPS) are expected to be new matrices for ACM, in order to utilise their high impact strength and weatherability.

Toho Rayon and BASF together have developed a blend fiber of PAN-based carbon fiber and PEEK. PEEK was first synthesised at ICI in 1977, and is employed widely for insulators in space shuttles and aeroplanes, as a coating material for electric wire used in nuclear reactors, oil exploiters and subways, and in a composite matrix together with carbon fiber or glass fiber. PPS is widely applied for use as valves of car exhausts, gears, electric switches, insulation parts, microwave ovens, steam irons, etc. Solvent resistance is important for the matrix resin used in the aeroplane to prevent corrosion by lubricants or other chemicals used for maintenance. Lightness and mechanical strength are not the only criteria. Other physical performances such as compressive strength, impact resilience and heat-resistance in the wet state are required for the primary structural material of aeroplanes. Since the aircraft industry is still small in Japan, CFRP is mainly used in the sports/leisure industry. For it to be employed in the general industrial area, including the car industry, further reduction in costs and improvement in its moulding/processing technology are essential. CFRP is moulded and processed conventionally using the same method as that developed for glass-fiber reinforced plastics. An efficient moulding/ processing technology needs to be developed for CFRP. If CFRP could be applied generally in car bodies, fuel costs would decrease because of reduction in weight. Recently, a CFRP body was employed in a Formula One racing car. Ford Motor Co. developed a 66 inch (1.67 m) (total length) CFRP drive-shaft of CF/glass/vinylester (20/40/40). CFRP is confidently expected to find more applications in the car industry following the lead of Ford. In 1981, Poli Motor Co. applied CFRP for the construction of a plastic car engine, which Ford adopted in its racing car. A considerable part of this car engine is made up of carbon fiber reinforced

Table 2.5. Amounts of carbon fiber reinforced plastic (CFRP) used in aeroplanes over a 13 year period

Model	First flight	Assumed weight of structural airframe (tonnes)	Weight of CRFP used (tonnes/aeroplane)	Amount of CRFP in aeroplane (%)	Weight of carbon fiber used (tonnes/aeroplane)
Boeing 767	1981	51	1.6	3	1.0
Boeing 777	1994	84	13	15	8.4

plastics (CFRP). The 300 horse power Ford racing car (Rota T616), equipped with this plastic engine (see Fig. 2.9) completed the whole course of the Camel endurance race with an average motor revolution of 8,500–9,000 rpm. This success encourages the application of CFRP to other comparable fields.

Since the call for CFRP in the sports/leisure industry will be small, a much wider range of industrial applications for carbon fiber in composites is required to make the best use of its high performance, light weight and high functionality. The preparation of a super-conducting fiber is being investigated in Germany and Japan, by forming niobium nitrate on to the carbon fiber. Carbon fiber loudspeakers are already popular, in order to capitalise on the fiber's high rigidity and damping effect. If a carbon fiber cassette were used instead of a conventional aluminium cassette, the X-ray dose required to take a picture would be reduced to one-fifth, because of the better X-ray transmission of carbon fiber. CFRP was the material used for the antenna of the Halley's Comet space probe MSIT5, to exploit its size stability against the temperature change and electrical conductivity. Research on CFRP is also now more active in Europe. Toray and Toho Rayon supply the know-how for the production precursor and refractory technology to Sofical Co., France, and Enka Co., Germany, respectively. Similar projects are also going on in the USA which could be extended to Europe, but on a smaller scale, such as in the European Airbus project and the Hermes project to launch a European space shuttle. According to A. R. Buncel, the President of the European Society of Composite Materials, there is need for CFRP to expand further in the field of aircraft technology, since it could solve the problem of metal fatigue. The composite content of European helicopters was increased to 35% by 1990. The third US–Japan Composite Materials Conference held in Tokyo in 1986 was followed with keen interest all over the world and developments have been rapid since. The space/aircraft industry in USA is military-oriented, and most of the achievements in this field are not published. Nevertheless, it is evident that the USA certainly leads in the science and technology of

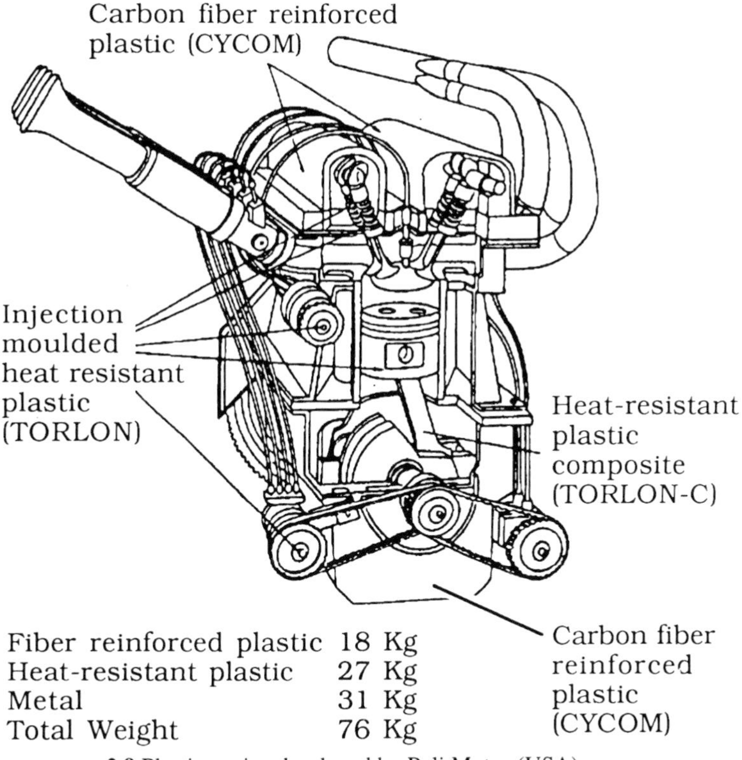

2.9 Plastic engine developed by Poli Motor (USA).

composite materials, and the technological gap between USA and other countries remains substantial.

2.11 High-technology boats of carbon fiber

Mitsubishi Rayon Co. joined the research/development project organised by the "Active Technology Promotion Council of Hiroshima", within the programme "Local Technology Promotion Activity" sponsored by the Small and Medium Enterprises Agency of Japan. This activity led to the production of a high-performance sea-worthy cruiser (Fig. 2.10) made of the carbon fiber composite material Pyrofil; Hiroshima Industrial Technology Center also cooperated. This cruiser was the first example of the application of carbon fiber technology developed on the basis of the plastics, fiber and textile production-processing know-how of industries in the Hiroshima area. With its strong technological base in producing and processing carbon fiber composite materials, Mitsubishi Rayon Co. joined this Promotion Council in 1983 and developed carbon fiber cloth suitable for cruisers, in cooperation with

2.10 Carbon fiber boat (Mitsubishi Rayon Co.).

Chugoku Spinning and Weaving Co. (Fukuyamashi, Hiroshima, Japan), which has extensive technological experience in fiber weaving, including carbon fibers.

The high-performance cruiser was designed by Blue's Naval Design (Hiroshima), which specialises in the design of power boats and cruisers, and was built by Kazenoko Boat Yard (Kurahashi Island, Hiroshima) during 1986–87. The cruiser is 16 gross tonnes, has 12 crew, is 15 m long and 4.4 m wide, and was produced from 500 m^2 carbon fiber cloth of 200 kg. The cruiser is operational and no problems were encountered in its commissioning. Mitsubishi Rayon subsequently expressed their intention to develop the application of carbon fiber composite materials for ship/marine technology. The carbon fiber cloth is used for structural parts, such as the engine mount, partitions and the deck, so that its total weight is reduced by 30% in comparison with a conventional cruiser made from FRP. Its cruising speed is 35 knots (65 km/h). This experience has demonstrated the considerable influence of carbon fiber cloth on energy conservation and vibration reduction, and has also proved its safety and durability.

2.12 Final stage of pitch-based carbon fiber development

Pitch-based carbon fiber is one of the most interesting new materials at present. In Japan, 23 companies including energy-related industries (such as oil companies and steel industries consuming a large quantity of coal) and fiber/textile industries are investigating the production technology of pitch-based carbon fiber on a commercial basis, which is now almost in its final stage of implementation. Although the price of PAN-based carbon fiber has been reduced considerably in response to the large demand from the ACM industries in the USA, it is still rather expensive. In contrast, tar or pitch is cheap as a by-product from oil- or coal-processing. The carbonising efficiency in carbon fiber production is approximately 50% in PAN-based carbon fiber, whereas it is as high as 80% for pitch-based carbon fiber. Thus, the production cost of pitch-based carbon fiber could be considerably less than PAN-based carbon fiber, based on original materials and energy conservation considerations. If its price could be reduced, its area of application would be considerably expanded especially in the car industry, where it could have a big market. Oil and steel industries could also benefit substantially by venturing into the new materials field using pitch-based carbon fiber. The companies producing PAN-based carbon fiber and production in 1996 are shown in Fig. 2.11.

The development of pitch-based carbon fiber is led by Union Carbide Corp., USA, followed by Amoco Chemical Co., and Kureha Chemical Ind. Co., Japan. Pitch-based carbon fiber can be classified on the following:

1 The HP (high-performance) grade of long staple prepared from the optically anisotropic mesophase pitch.
2 The GP (general purpose) grade of short staple prepared from isotropic pitch having no fiber structure.

Union Carbide produced *ca.* 200 tonnes filament (continuous fiber) and *ca.* 100 tonnes staple annually (in 1981), and Kureha expanded its capacity to 900 tonnes staple/year in 1986. Two types of the Union Carbide product are available with modulus of 390 and 530 Gpa and a 700 GPa type is also available for trial purposes. The Kureha product is a GP grade with modulus of *ca.* 40 GPa and tenacity of *ca.* 0.8 GPa. A new company, Donac, was established in Osaka in 1988 jointly by Osaka Gas Co. (40%), Dainippon Ink & Chem. Inc., (40%) and Nippon Sheet Glass Co. (20%), which set up a production line for 300 tonnes per annum pitch-based carbon fiber of GP grade. Curled Donac carbon fiber is produced by eddy spinning, whereas straight Kureha carbon fiber is centrifugally spun. Donac-produced carbon fiber will be sold through the three parent companies, which will develop independently the unique applications of pitch-based carbon fiber according to their own special interests and experience. These are composites using plastics (Osaka Gas), resin-impregnated carbon fiber mats (Dainippon Ink & Chem.) and blended yarns (Nippon Sheet Glass). Mitsubishi Chemical Industries built a production plant with an annual capacity of 500 tonnes for

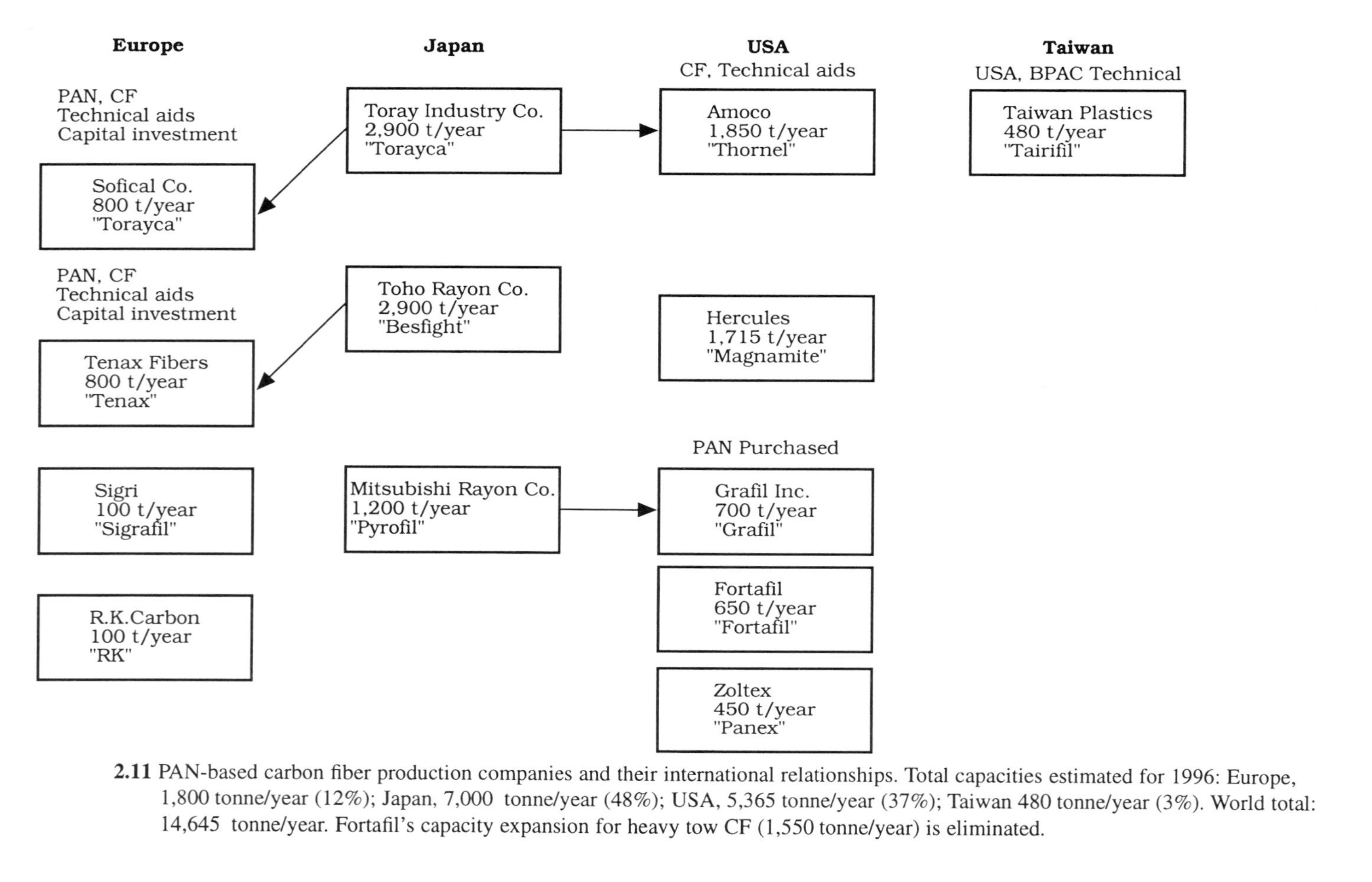

2.11 PAN-based carbon fiber production companies and their international relationships. Total capacities estimated for 1996: Europe, 1,800 tonne/year (12%); Japan, 7,000 tonne/year (48%); USA, 5,365 tonne/year (37%); Taiwan 480 tonne/year (3%). World total: 14,645 tonne/year. Fortafil's capacity expansion for heavy tow CF (1,550 tonne/year) is eliminated.

carbon fiber filament at its Sakaide Factory in 1987. Mitsubishi decided that coal-pitch based carbon fiber Dialead was more profitable than PAN-based carbon fiber, from the standpoint of the original material used and the physical properties that could be developed over the longer term.

There are still many difficulties to be overcome to establish the necessary industrial production technology for the less expensive pitch-based carbon fiber of the HP grade. The first problem is how to secure cheap pitch of constant quality for spinning. The improvement in the mechanical strength of synthetic fibers was mainly due to the purity of the original materials. Pitch or tar is a complex non-homogeneous material, so it is unsuitable as the basis for further syntheses. How should this paradoxical situation be coped with? Secondly, the viscosity of pitch or tar is sensitive to temperature differences at the spinnerets. Temperature control is, therefore, vital to yield carbon fiber of constant quality. Since the fiber immediately after spinning is extremely weak, it is a difficult practical problem to adapt the fiber to the sintering process. Finally, can pitch-based carbon fiber of the HP grade be supplied less expensively? The cost of pitch-based carbon fiber of the GP grade can be reduced by increasing the conversion efficiency, since the processing of the original pitch is relatively easy and the original material is cheap. In the HP grade, a long controlled process is required to prepare the pitch mesophase, and consequently the price of pitch-based carbon fiber of HP grade cannot be reduced dramatically. Here, the preparation of the pitch mesophase of homogeneous quality is important to produce carbon fiber of high strength. The pitch-based carbon fiber crystal structure is determined by the graphite structure, which in turn determines the physical characteristics of pitch-based carbon fiber. Thus, the modulus of pitch-based carbon fiber is larger than that of PAN-based carbon fiber, but the strength of pitch-based carbon fiber, influenced by structural defects, cannot achieve that of PAN-based carbon fiber. The future of pitch-based carbon fiber depends on strength.

Recently, Nippon Steel Corp. and Mitsubishi Oil Co. have opened up a new field of application for pitch-based carbon fiber by adding new characteristics to their conventional products. They developed a carbon fiber filament of the GP grade to improve its performance in comparison to a conventional pitch-based carbon fiber staple. This new type of pitch-based carbon fiber can be placed between the HP and GP grades in terms of its performance and price. Pitch-based carbon fiber staple of the GP grade was first industrially applied with CFRC (carbon fiber reinforced cement). Kajima Corp. employed CFRC as a curtain wall for the 37 storey Arcs-Hill Mori Building in Akasaka, Tokyo. Some 160 tonnes of pitch-based carbon fiber of the GP grade were used for this building, and consequently 4,000 tonnes of steel-frame were not required. This reduced the weight of the external walls by 60% and the earthquake load by 12%. Karma Corp. plans to use CFRC curtain walls for its second Main Office Building, and the Akasaka Dai-Ichi Mutual Life Insurance Co./Kajima Common Building. This field is discussed in more detail in a subsequent chapter.

Pitch-based carbon fiber of the GP grade is applied as the reinforcing material to industrial plastics such as PP (polypropylene), polyacetal and fluoroethylene resin, in order to improve their heat resistance, wear resistance, size stability, lubricity, chemical resistance and conductivity. It is also used for the insulators of incinerators, the moving parts of large cars, as C/C (carbon/carbon) fiber composites for electrodes, and for the brakes of aeroplanes. Carbon fiber is expected soon to replace asbestos fully for producing the friction brakes of cars.

Carbon fibers experienced an improvement of their mechanical properties during the 1980s which is unparalleled in the history of man-made fibers. The progress subsequently has been more of a specialised nature, where high modulus and strength types were further developed to ultra-high modulus and strength types from the early 1990s. Figure 2.12 summarises this overall development with the carbon fiber Tenax produced by the Toho Rayon Company.

In 1995, the consumption of carbon fibers had increased to 9,000 tonnes. Briefly, in 1990, the demand stagnated somewhat due to the reduction of military budgets in the USA. A special boost was given by the earthquakes in California in the early '90s, since carbon fiber reinforced bridge pillars survived the Los Angeles earthquake of 1993. This was an enormous trigger for civil engineering application, and subsequently in 1995 the earthquakes in

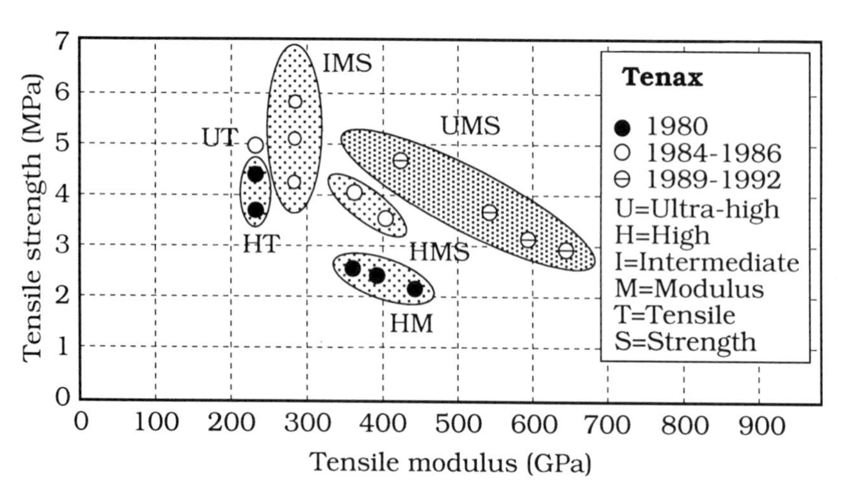

2.12 Development carbon fibers (PAN) 1980–95.

the Kobe area of Japan initiated further usage in the construction and building sector. The demand is expected to rise to 12 500 tonnes in the year 2000.

2.13 The future of super-fibers

Super-fibers are now widely applied in the field of aerospace technology, automobile technology, ocean developments, optical networks, and the leisure industry, and for the production of materials such as safety/protective garments and frictional materials. These applications are mainly to replace metal parts and to reduce the weight for the purpose of energy conservation. Therefore, the development of high-tenacity fiber materials of light weight is urgently required.

There are several problems to be solved in order to apply super-fibers in the clothing area. Super-fibers elongate by *ca.* 3–4%, which reduces their potential for garment manufacture. Their wear comfort is not particularly good either, nor does the material dye substantively. Since the less amorphous regions are most available for dyeing in the liquid crystal spun fibers, super-fibers can be dyed by mixing with pigments, which restricts the colour range and colour brilliance. Super-fibers are also too expensive to be used in the fashion industry, so at present they must find their application in the non-clothing area, by capitalising on their unique physical properties. Thus the future prospects for super-fiber depends on further high-performance improvement, in terms of tenacity and modulus.

The limiting values of tenacity and modulus can be theoretically estimated for each synthetic fiber. Although the observed values vary with the method of measurement, the modulus and tenacity of aramid fiber (Kevlar 49) can reach 90% and 17% of the respective theoretical values. In the case of less rigid polymers such as polyethylene, the maximum tenacity so far achieved is about 10% of the theoretical limiting value. The tenacity difference between the real and theoretical values is still large, and many researchers are competing to reduce this gap and produce fibers that ultimately possess modulus and tenacity close to the limiting values theoretically predicted. There is competition to achieve this, the result of which will influence an unexpectedly wide area of science and technology. New materials will be developed further to add not only high performance, but also high functionality. ACM and molecular composites are also expected to make active progress, particularly in this latter area. Molecular composites are particularly interesting, since their reinforcing effect will increase by micronising reinforcing materials from macroscopic fibers to molecularly dispersed rigid macromolecules. A molecular composite of rigid and flexible macromolecules (developed by M. Takayanagi, former Professor at Kyushu University) exhibited the highest

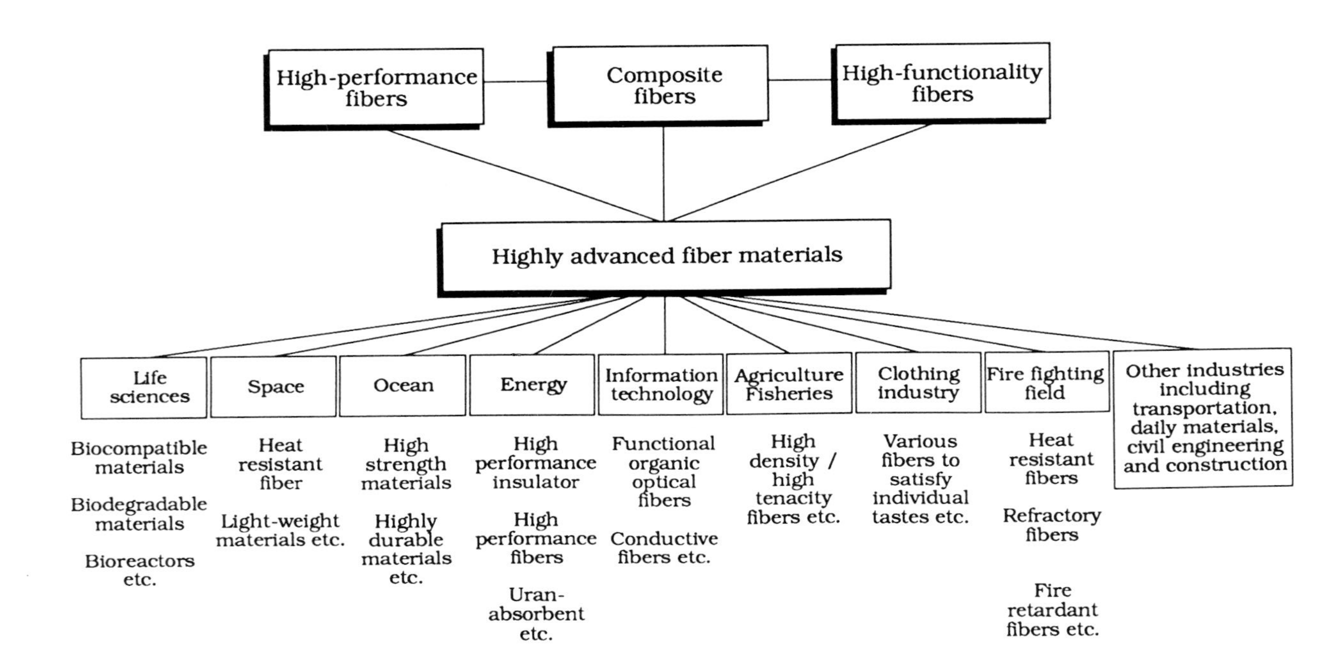

2.13 Development of advanced fiber materials.

heat resistance (stable at 500 °C) yet found among organic materials. Professor Huang at Deighton University has also developed a molecular composite which has a better performance than aluminium.

In the area of fibers, nature still has a great deal to teach us. A fiber consists of crystalline and amorphous regions, but the true function of the amorphous region in natural fibers, in particular, is still not clear. The details of how a silkworm undertakes its spinning with perfect energy conservation still remains a mystery. This points to the fact that we should learn more about the basic functions of natural fibers in developing synthetic fibers for new functions and for human clothing. The demand for high-performance and high-functionality fiber materials will increase to support comfortable human activity. If the "ultra-super-fiber" and its related applications technology could be established, the emerging industrial structure could influence the social environment in a major way (Fig. 2.13). In order to design specific fiber functions and so produce the ideal fiber, a new system of fiber science must be set up. Basic research must support these developments. It is now becoming clear that a study of natural polymers can lead us to new high-functionality polymers. For example, a knowledge of natural polymers is now leading to the development of new biocompatible polymers.

A better and more effective industrial–academic–government cooperation system is needed in Japan, Europe and the USA to enable this technology to develop internationally. The higher education systems must also train more researchers in the field of fiber science.

The fiber has a huge potential and has as yet unknown functions that must be exploited and applied in the service of humans.

3 High-touch fibers

Among speciality fibers, the "high-touch fiber" is defined as a fiber that especially appeals to human sensitivity, such as the senses of touch, sight and hearing. The touch of a fiber is determined by its mechanical strength, bulk, warmth, softness, smoothness and/or stickiness of the material. Visually, the colour arrangement, hue, colour tone, lustre, shape, pattern and bulk are important factors. Silk-scrooping is pleasant to the ear. The finesse or texture of the whole determines the quality of the clothes.

Figure 3.1 shows the relation between the cloth-texture and the five human senses. In the first illustration, the material is improved to give a good hand-touch feeling and texture. The example includes hygroscopic polyester and nylon yielding the cool impression given by cotton, acrylic textiles, and polyester fibers, and having a better quality than silk. The second example adds the visual characteristics in order to give an unexpected impression. The third example shows the fibers with new functions such as the ceramic-blended fibers with high heat-insulating characteristics to counter the deep penetrating effect of far infra-red radiation. Better conducting fibers are used for dust-proof clothes in the electronic industry where high precision is required. The fourth example is clothes for fun, represented by scented textiles and optical and polarising fabrics that change colour according to temperature or the visual angle, so polarising light. The following sections will illustrate representative examples of these high-touch fibers.

3.1 A silk-like fiber that surpasses natural silk

Silk is distinguished by its characteristic lustre, vivid colouring, puffiness, draping and silk-scrooping. A silk-like fiber is defined as a chemical fiber that generally possesses at least some of these silk characteristics, or that which specifically provides characteristics of silk as a whole.

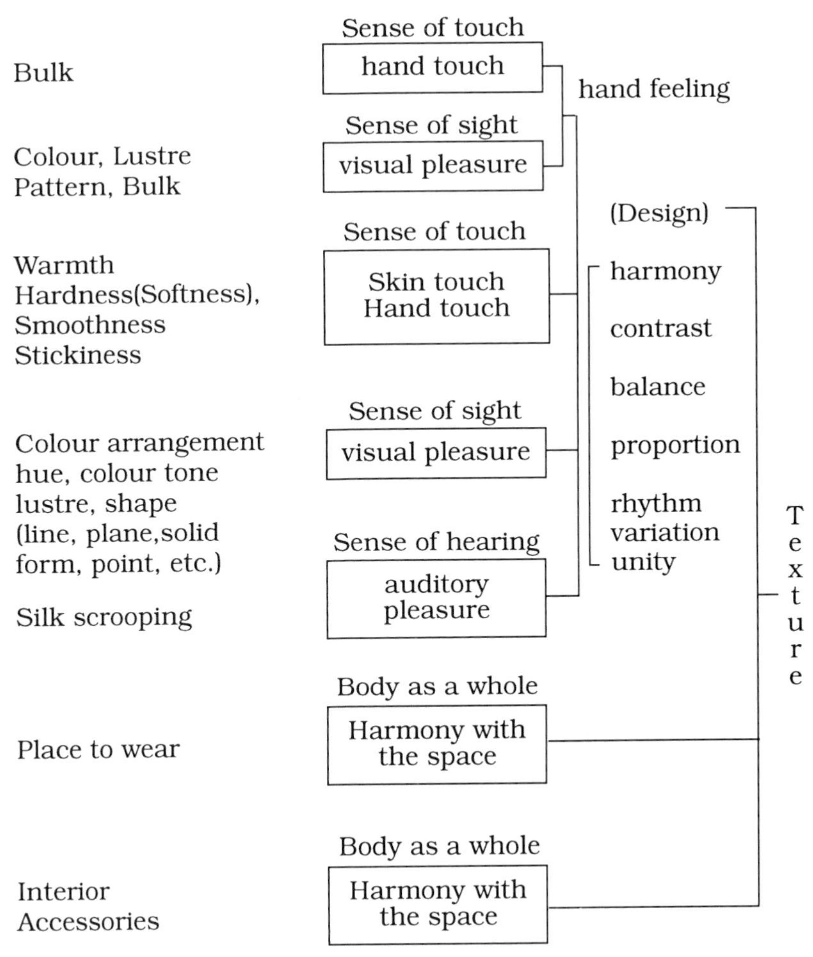

3.1 Cloth texture and the five senses.

The production of artificial silk has long been attempted. The first artificial silk chemical fiber started its production in 1891, following Count Chardonnet's invention of a silk-like fiber (rayon) derived from nitrocellulose. In 1938, Carothers of Du Pont developed the polyamide fiber "nylon" from an investigation designed to synthesise the amide-link present in silk. Nylon was an epoch-making fiber with silk-like characteristics, and even now is widely used as one of three big synthetic fibers, which include polyester and acrylic fibers. These two inventions, although on a different time-scale, were motivated by an affection for silk, and produced completely new types of fiber.

The commercial production of synthetic fibers in Japan started in 1951 with vinylon, followed by nylon in 1952, acrylic fibers in 1957 and polyesters in

Table 3.1. Development of silky polyesters: history and technology

Year	Technology developed	Aim	Effect
1960	Triangular cross-section	Silk-like cross-section	Silk-lustre, silk-crispness
1970	Alkali treatment	Sericin-removal imitation	Drape
1971	Multiple fibers	Silk fineness (1 denier)	Fineness, elegance
1975	Blended fibers of different shrinkage	Silk puffiness	Silk softness, puffiness
1976	Change of surface structure	Imitation of wild silk surface	Silk-scrooping
1985	Unevenness	Silk unevenness	Natural silk-like

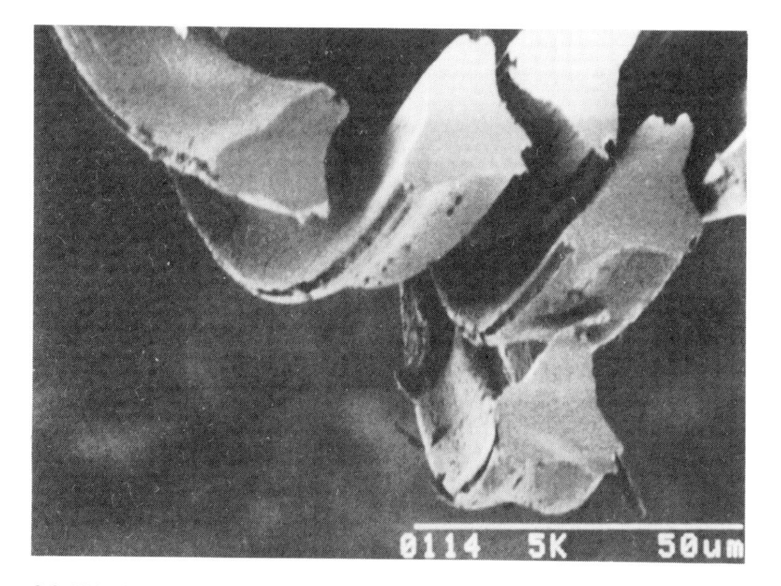

3.2 Sillook Royal S (Toray Industry Co.).

1958. These synthetic fibers have a round cross-section, and give a flat and paper-like touch. Since natural silk has a triangular cross-section, each company competed to produce fibers with a triangular or non-circular cross-section. Indeed, most of the textile technologies available today were established in the course of developing silk-like materials.

The history of silk-like material development can be divided into four or five generations, as analysed in Table 3.1 in terms of the development concept and the technical characteristics. The production of cleaner polymer and triangular cross-sectional fibers enabled an increase in transparency, lustre and crispness in the first generation to approach natural silk. Silk-like puffiness was achieved by varying the mono-filament length in the second

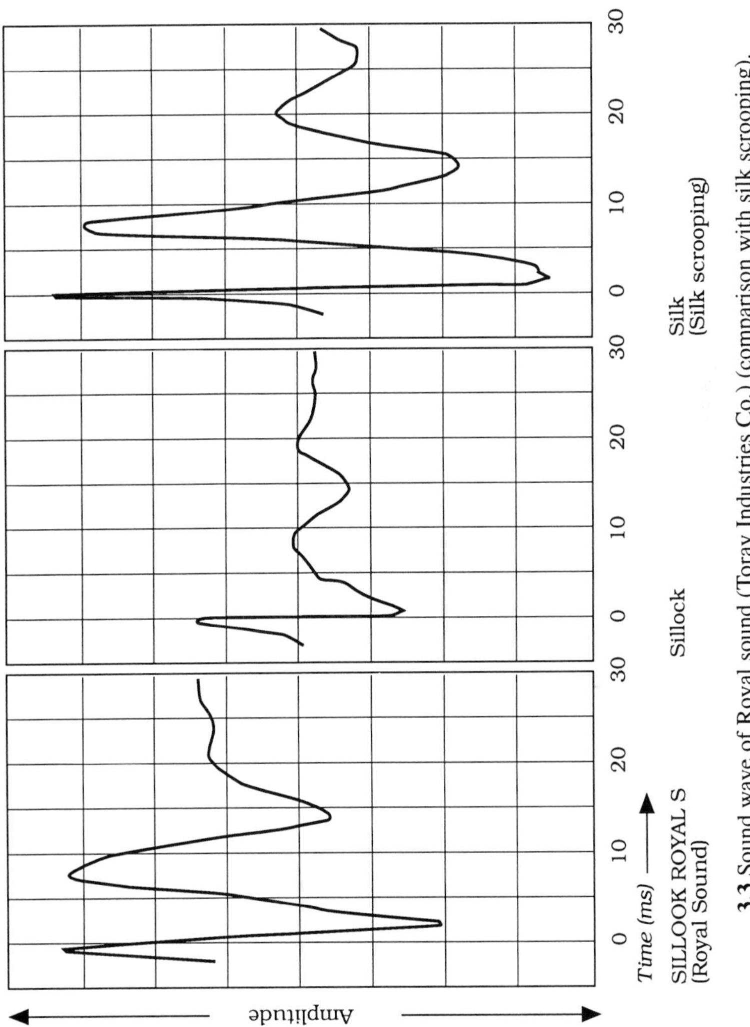

3.3 Sound wave of Royal sound (Toray Industries Co.) (comparison with silk scrooping).

generation. Combined temporary twisting and irregular drawing techniques were adopted in the third generation to gain the unevenness of natural silk and produce spun silk. Microscopic technology was applied in the wave of high technology to develop the three-petal cross-section (see Fig. 3.2) or irregular cross-section technology to produce ultra-fine fibers and imitate the silk-scrooping sound.

Toray developed the silky polyester fiber Sillook Royal S, which has a three-petal shaped cross-section with micro-slits of the order of 0.1 mm at each petal top. These micro-slits absorb the reflected light, and provide the vivid deep colour and elegant anisotropic lustre simultaneously. The micro-slits effectively prevent the worn-out shaping of the cloth, and introduce the elegant silk-scrooping and the pleasant cloth-rustling sound when two edges of a micro-slit are touched and rubbed. Sillook Royal is silk-like, not only from its hand touch, but also from a visual and auditory standpoint.

Cloth rustling and silk scrooping are generated by rubbing clothes or fibers against each other. Here the rubbing operation gives the rustling sound (high-frequency sound) by high-speed rubbing and scrooping (low-frequency sound) by low-speed rubbing (graphically shown in Fig. 3.3). The silk-like materials generated over these years were developed, not only to imitate natural silk, but also to create a new hand touch and elegance that only synthetic fibers can provide. For example, new technology was developed to

3.4 Sillook-Sildew (Toray Industry Co.).

spin ultra-fine fibers, to form slits on the fiber surface, to spin the fibers with uneven cross-section (thick and thin in parts), etc. and in an unexpected way created a new type of elegance quite different from natural fibers. Toray developed the conjugate yarn with different shrinking characteristics (Sillook-Sildew) which consists of high- and low-shrinking fibers. When the fabric or knit product of this yarn is heated after the dyeing process, the high-shrinking fibers shrink and relax the low-shrinking fibers. As a result, the fabric becomes puffy and yields a new hand touch different from natural silk (see Fig. 3.4). Although the molecular design of polymers and fibers will be re-evaluated from a stand-point different from that mentioned above, the fabrication of silk-like materials is an evergreen subject for textile engineers.

3.2 The challenge of ultra-fine fibers

Several new microfibric technologies have been developed in Japan, represented by the "sea–island (islands-in-the-sea)" type (Type 1), the "separation" type (Type 2), the "fiber-splitting" type (Types 3 and 4) and the "multi-layer" type (Type 5) as schematically shown in Fig. 3.5. Type 1 shown here is somewhat different from the "sea–island types" currently employed by Toray and Kuraray, and even for these two companies production techniques are different. In this first classification the fiber is composed of two component polymers A (sea) and B (cores or islands). Toray arranges them in parallel, with polymers A and B in the conjugate spinning nozzle prior to spinning-out, but Kuraray blends them randomly in the extruder. When spun and fabricated, the A component (polystyrene) is removed by dissolving in a solvent to produce ultra-fine fibers of the B component (polyester or polyamide by Toray and nylon by Kuraray). Although the initial mono-filament (AB composite fiber) is rather thick (3–5 denier), the mono-filament of the B component is extremely fine (less than 0.1 denier). Owing to the difference between the processes, the microfiber of Toray is continuous and homogeneous in terms of its thickness (0.1 denier), whereas Kuraray's varies from 0.01 to 0.1 denier discontinuously but can be processed in the reversed "sea–island" to produce the lotus-root-shaped (porous) fiber (see Fig. 3.6). These different "sea–island" type fibers are commercialised for artificial leather such as Ecsaine (Toray, known as Alcantara in Europe and Ultrasuede in the USA) and Kurarino (Kuraray). Type 2 (the "separation type") can be considered as a variation of the "sea–island" type or the "fiber-splitting" type. The recent developments in this area are described in Chapter 7. Types 3 and 4 in Fig. 3.5 are termed the fiber-splitting" type, whereas polyester and nylon are composite-spun and then split into their respective components. Type 3 is a

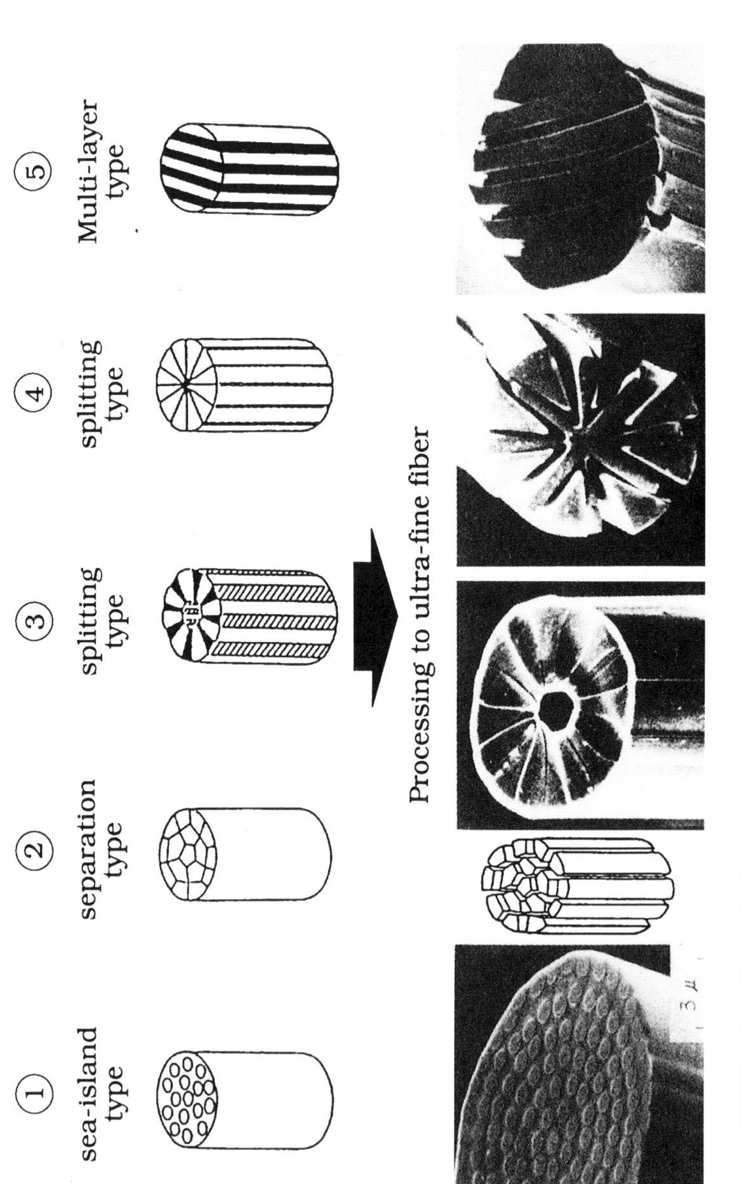

3.5 Ultra-fine fiber technology by composite spinning.

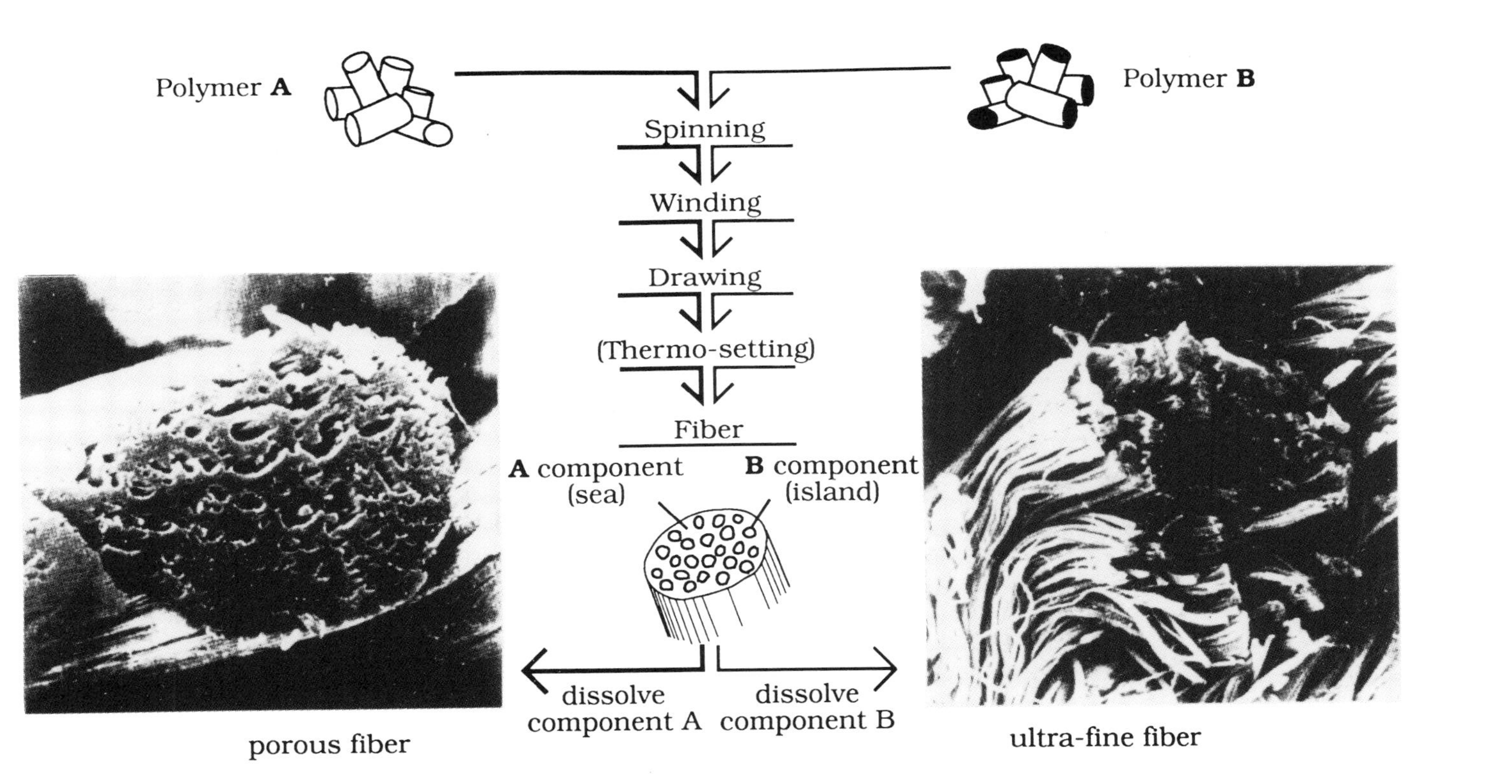

3.6 Fiber formation and shape (Kuraray Co.).

hollow composite fiber of polyester and nylon developed by Teijin (commercially available as Highlake and Elec). The composite fiber is split mechanically into 16 sections, and its mono-filament after splitting has a homogeneous cross-section of 0.23 denier. Kanebo developed Type 4 microfiber with nylon in the core and polyester (commercially available as Belleseime and Savina), and its mono-filament after splitting is on average 0.15 denier thick.

Melt-blow spinning, tuck spinning, super-drawing, etc. are also employed to produce microfibers. These microfibers are mostly applied to the hair fabrics such as non-woven (artificial leather) or knit fabrics and high-density fabrics for special functions such as water repellency. Based on the high technology of microfibers, each company in its own way is looking for a market for new types of high-touch fibers with specific characteristics. Microfibers have opened up a new field of applications and expanded the expression potential of fibric materials.

Although ultra-fine fiber technology was developed first to produce artificial leather, its know-how was further developed to produce even finer fibers, which found an unexpected application as a wiping cloth for spectacles. The cloth can also be applied to clean car mirrors, remove oily dirt from windows, clean computers and electro-optics devices, polish and clean jewels and noble metals, polish crystals, clean showcases and frames, clean panels of audio/visual devices and optical discs, remove fingerprints on photos and films, and clean plastic products, etc. Here high-density fabric is variously processed into puffs, gloves, quilts and even into tapes, according to the handling convenience needed. For example, high-density fabric tape is used as a built-in type cleaner for machines. High-density fabric is thus a new frontier fabric, which may find even wider applications.

3.2.1 High-touch material Zepyr 200

Kanebo developed the conjugate yarns Belima and Belima X and an artificial leather/suede Belleseime by imitating the bicomponent structure of silk or wool. Belima and Belima X are produced from the radial conjugate fibers of multi-layer bicomponent filament by splitting into ultra-fine fibers of 0.1–0.5 denier. The radial conjugate fibers were further developed to a highly soft-touch nylon fabric Zepyr 200. The conjugate fiber for Zepyr 200 is composed of 75% nylon (0.67 denier) in eight triangular segments and 25% polyester (0.17 denier) in a radial segment. When the woven or knitted fabrics of the radial conjugate fiber are alkali-treated, the polyester portion is removed and the fiber is split (Fig. 3.7). Then the fabrics become 100% nylon, and they can

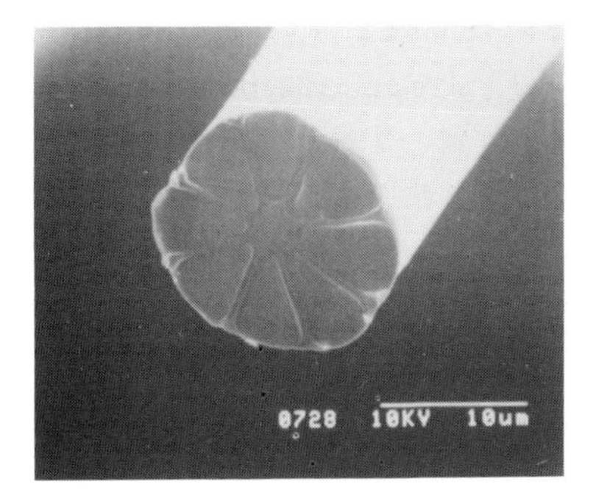

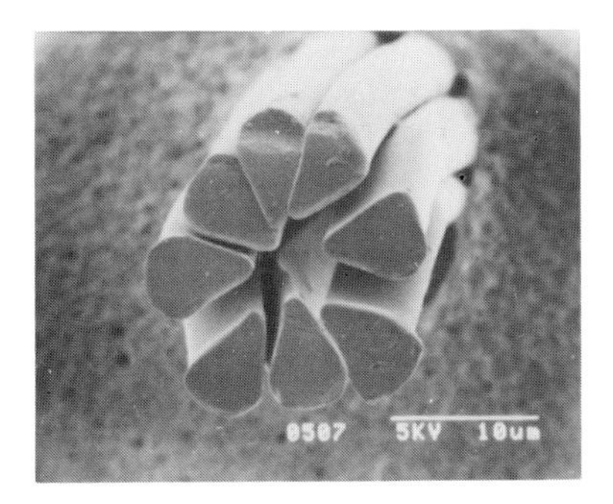

3.7 Electron microscopic view of Zepyr 200 cross-section: top, before splitting; bottom, after splitting (Kanebo Ltd).

be finished by the conventional nylon processing to Zepyr 200, which has an extremely soft touch and an elegant appearance. For example, when the original yarn of 50 filaments, 100 denier in total, is alkali-treated, the resulted yarn is then composed of 400 filaments, 75 denier in total, i.e. each filament is extremely fine nylon of 0.18 denier. The cross-section of the fiber shows fine flower petals of several micrometres as displayed in Fig. 3.7. Because of its characteristic cross-section shape, Zepyr 200 is highly bulky, extremely soft and brilliantly coloured. Since nylon is lower in Young's modulus and density than polyester, nylon is much softer and lighter than polyester of the same fineness. Thus Zepyr 200 is a real new fabric which offers the following:

1 A silk-like soft touch and rich drapability without the rustling noise of polyester fabrics.
2 A mild and decent colour of brilliance without a see-through effect (owing to the characteristics of an assembly of eight petal-like cross-sections).

The extremely fine denier technology accumulated in developing artificial leather has thus been applied to general fabrics for clothing in the development of Zepyr 200, which pioneered a new era of nylon which expanded its application to the field of ultra-fine fiber fabrics for high-quality garments and sportswear.

3.2.2 Perfect cleaning with ultra-fine fibers

Dirt or smudges on spectacles are always a nuisance to those who wear them. Such dirt can spread over the entire glass surface when wiped with a handkerchief, or the glass may be scratched when wiped with a tissue paper. As mentioned previously, a special wiping cloth for spectacles was developed by utilising high-density fabrics of ultra-fine fibers, which now sells by the million.

Kanebo launched the first spectacles-wiping cloth made of ultra-fine fiber fabric (Belleseime) in 1978. Kanebo developed its first ultra-fine fiber Belinu in 1974, and then Belima X in 1975. These fibers are composed of polyester radially divided into eight sections by polyamide (Fig. 3.5, Type 4), and split in 8–13 mono-filaments by a technology that involves temporary twisting or swelling. The basic ultra-fine fiber cloth is knit, with each mono-filament after splitting 0.1–0.2 denier thick.

Teijin developed the hollow splitting-type composite fiber utilised for spectacles-wiping cloth in 1982. This ultra-fine fiber Highlake is also made of polyester and polyamide, radially sectioned and arranged around a hollow fiber (Fig. 3.5 Type 3). The original fiber is split into 16 mono-filaments, with each component mechanically woven into cloth. Each mono-filament is 0.23 denier thick.

Kuraray adopted a multi-layer type (Fig. 3.5, Type 5), where nylon and polyester form 11 layers alternatively. The composite fiber is longitudinally split into 11 flat ultra-fine mono-filaments (0.2–0.3 denier each) in the dyeing process, after being woven into cloth. Although the ultra-fine effect is less apparent in comparison with other types, the composite fiber of Type 5 can be split easily, and the cloth of this composite fiber is soft because of the extremely flat shape of filaments (1–2 μm thick and 10-15 μm wide). Nylon shrinks during the splitting process to add extra soft handling to the cloth.

Toray started research and development in 1976 into the ultra-fine fiber cloth UT-C and the wiping cloth Toraysee, applying the technology developed for the "sea–island" type composite fiber. As described, these have numerous cores and were developed for the non-woven artificial leather Ecsaine in 1970.

Each company developed its own composite-spinning know-how for producing ultra-fine fibers and as a result the respective products have their own characteristics. Here, the cleaning mechanism of the ultra-fine fiber wiping cloth will be demonstrated by using as an example the Toraysee series developed by Toray. Most of the sticky dirt is caused by dust accumulating on thin layers of fat, which merely spreads and is barely touched by conventional wiping cloths, such as the chamois leather (natural leather) or waste cloth. This is because the fiber of these wiping cloths is normally 10 mm thick and is unable to capture the 1 μm thick oil layers. Toray offers many types of Ecsaine (Alcantara, Ultrasuede); a standard type is made from a composite fiber with 16 core components and a soft-thin type made from composite fiber with 36 core components. They developed the composite fiber with 70 core components for the Toraysee and UT-C series. Since 10 sea–island type composite filaments are twisted into a fiber and woven, a single fiber of the fabric consists of 700 micromono-filaments after the seam components have been removed by dissolving in a solvent. The filament density of the resulting fabric is, therefore, approximately 220,000 filaments per inch. A mono-filament of this fabric is only 0.05 denier or 2.0 μm thick, and is claimed to be the finest continuous filament commercially available in the world. These ultra-fine filaments can penetrate into the thin fatty layer of dirt and trap it within the micro-pockets among the filaments. Another reason for the excellent wiping effect is because of the good compatibility of the Toraysee series (made of polyester) with skin fat (fatty acid ester). The dirt trapped in the micro-pockets can be removed and the wiping characteristics can be regenerated by washing.

The wiping mechanism of the Toraysee series can be attributed to the following effects according to Dr Okamoto, Toray Industry Inc.:

1 *Sharp-shaving effect.* The fatty layer of dirt attached to the spectacle surface is about 1 μm thick. When wiped with a conventional cloth with thick filaments, the fatty dirt is caught and stays on the surface of the filaments. Thus, the cleaning effect deteriorates considerably when the whole surface

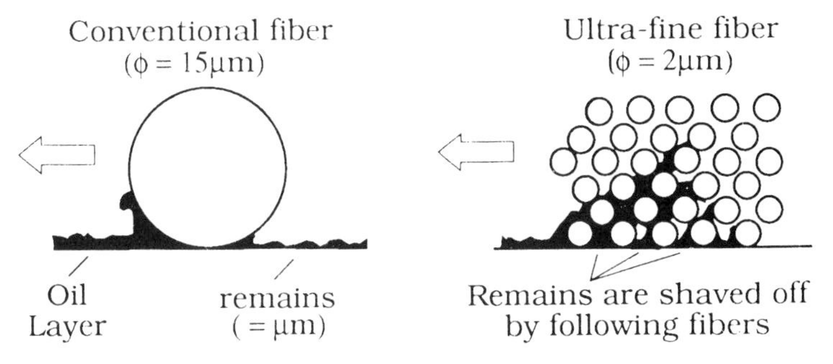

3.8 Multi- and sharp-shaving effect.

is covered. When the wiping cloth is made of ultra-fine fibers, the fatty dirts are removed and drawn into the micro-pockets (see Fig. 3.8).

2 *Multi-shaving effect.* The contact frequency with the glass surface is much higher for ultra-fine fibers than for a single fiber before it is split into 70 ultra-fine fibers. As a two-blade razor is more effective in shaving than a single-blade razor, so a wiping cloth of ultra-fine fibers removes dirt more effectively than a single fiber. The ultra-fine fibers are thin and soft, and no scratch is left on the glass surface (see Fig. 3.8).

3 *Wide-contact effect.* A wiping cloth of ultra-fine fibers contacts over a wider area with the glass surface than a conventional wiping cloth. Thus the effective wiping area increases in comparison (see Fig. 3.9).

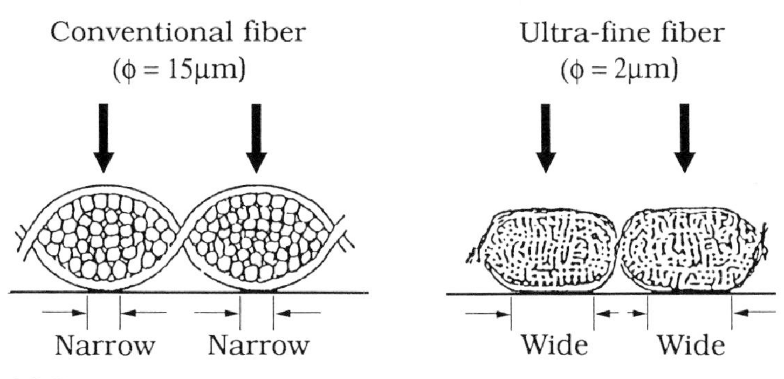

3.9 Wide-contact effect.

4 *Inner trap effect.* A wiping cloth made of ultra-fine fiber contains micro-pockets. As the cloth is pressed by the finger wiping the spectacles, the fatty dirt on the glass is adsorbed by the micro-pockets. When the pressure is released, the fatty dirt migrates into the inner part of the cloth where the fiber density is higher. After migration, there is less dirt on the surface area of the cloth, and the cleaning efficiency recovers (see Fig. 3.10).

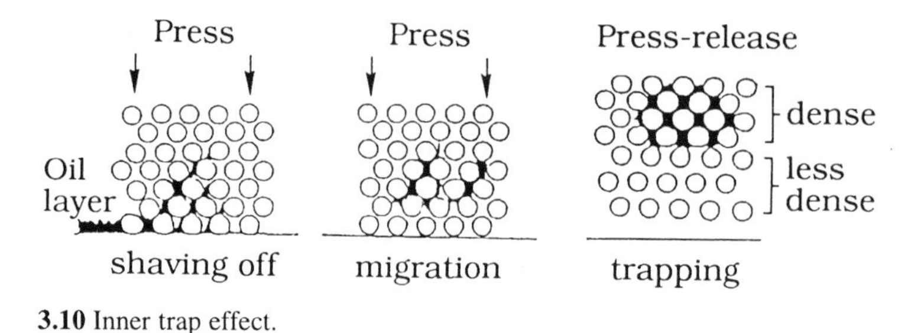

3.10 Inner trap effect.

3.3 The skin-like fabric Exceltech

The moisture-permeable/waterproof material Exceltech was developed by Unitika Ltd for use by the Chomolungma (Everest) expedition. The same material was also adapted for ski-wear and was used by many of the medalists in the downhill and slalom competitions at the Calgary Olympic Games in 1988, first and subsequently. This is because of its excellent moisture permeability and wide comfort range, confirmed by actual test conditions and in artificial weather chambers.

There are three main functions necessary for outdoor sportswear used in severe weather conditions. The first relates to physical mobility, and can be fulfilled by material elasticity, fitness and lightness. Safety and durability is a second requirement and must include mechanical strength, flame-resistance, colour fastness and dirt-resistance. The third function is body comfort involving absorption, thermal insulation, wind protection, waterproofing and sweat and moisture permeability. The first and second functions can be fulfilled by many conventional materials, but no suitable material available satisfied the demand for body comfort, particularly moisture permeability when sweating during exercise. The moisture-permeable/waterproof materials developed hitherto placed more emphasis on waterproofing and wind protection and these materials were not ideal for high-sweating conditions. They caused a sticky and stuffy feeling and even a decrease in body temperature due to dew condensation.

Exceltech utilises a polyaminoacid film that has the highest moisture permeability of all the synthesised polymer materials. Polyaminoacid is a type of protein-like polymer which is similar to molecular skin components. Exceltech is, therefore, a "skin-like" material that absorbs and releases water actively, according to environmental conditions. A technology was developed to modify and wet-coat polyaminoacid on to nylon or polyester cloth. The surface of Exceltech is porous ($\phi < 1$ μm) and these micropores are connected to each other to form microcells (Fig. 3.11). This microporous surface structure and molecular characteristics of polyaminoacid are the result of its α-helix structure, which as a result gives excellent porous waterproof characteristics. Consequently, the high moisture permeability, non-dew-condensation characteristics, thermal insulation and waterproofing required by windsurfers and ski-wearers are satisfied. Soft touch is an additional requirement for clothes for outdoor sports, and Exceltech satisfies all these requirements. It does not lose its soft feeling even below –20 °C, and is an ideal material for the clothes required for winter sports.

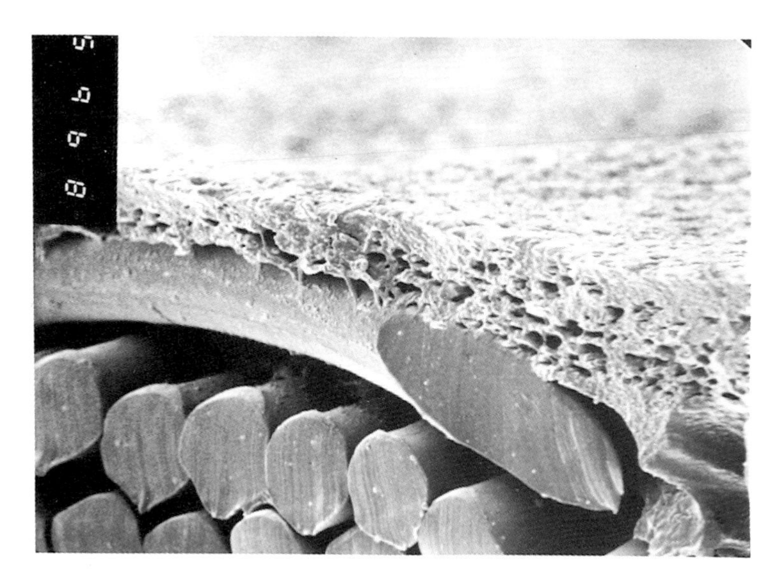

3.11 Cross-sectional view of polyaminoacid-coated Exceltech (Unitika Ltd).

3.4 Chameleonic fabrics

It has long been a dream to produce clothes in which the colour tone would change according to environmental conditions, such as weather or temperature. For example, the iridescent silk fabric in Japan achieves light dichroism by the way it is weaved. Only recently attempts were made to achieve a colour change in more fundamental way by applying the phenomenon of photochroism (colour change induced by light), thermochroism (colour change induced by heat) and solvation-chroism (colour change induced by moisture). Chameleonic fabrics have been developed by printing thermochromic and photochromic materials on to textiles, which thus change colour like a chameleon.

Although certain metal complex salts and cholesteric liquid crystal materials change colour reversibly with heat, these materials have both merits and demerits when used for clothing materials. Toray developed the thermochromic clothing material Sway by producing microcapsules (ϕ = 3–4 μm) to enclose heat-sensitive dyes, which are resin-coated homogeneously over the fabric surface. Here the microcapsules are dispersed homogeneously over the basic fabric (e.g. nylon) and coated with polyurethane resin. The microcapsule is made of glass, and contains the dyestuff, the chromophore agent (electron acceptor and the colour-neutraliser (alcohol, etc.)), which react and colour/decolour according to the temperature. Sway is a multi-

3.12 Ski-wear made of the thermochromic fiber Sway: top, at low temperature (below 11°C); bottom, at high temperature (above 19°C) (Toray Industries Co.).

colour fabric (basic 4 colours and combined 64 colours) which changes with temperature differences of over 5 °C over a temperature range between –40 °C and +80 °C. Here the colour changes thermoreversibly; for example, colour A to colourless (white) or as colour A to colour B as the temperature changes from low to high. At present, plain and printed clothes are commercially available. The changing of colour with temperature of these clothes is planned to match the applications such as for ski-wear (11–19 °C), ladies' clothing (13–22 °C) and for lamp shades (24–32 °C). Figure 3.12 demonstrates such Sway ski-wear.

3.5 Photochroism-controlled clothing material

American Cyanamide Co. first attempted to develop a photochromic material as a result of a contract from the US Army to synthesise the materials for combat uniforms to be used in the Vietnam War during the 1960s. Here photochroism denotes a colour change induced by light. Subsequently, there was no attempt to develop chameleonic fabrics using photochroism, until Kanebo Ltd developed Comic-relief in which the microcapsules enclosing the photochromic material are printed. This photochromic material is initially colourless, but colours from light blue to dark blue can be achieved by ultraviolet light of wavelengths 350–400 nm. The spiropiran-type organic compounds used for such photochromic material undergo photolysis and change colour by this ultraviolet irradiation. Although spiropiran compounds exhibit bright colours quickly in response to ultraviolet irradiation, they are chemically unstable and decompose by repeated exposure to ultraviolet irradiation. Kanebo employed a spiro-oxazine compound which is more stable

3.13 Comic-Relief (Kanebo Ltd).

than spiropiran itself. A T-shirt made of photochromic printed fabrics was introduced to the market in 1989 (see Fig. 3.13).

A handkerchief is available that changes colour when wet. Here the printed colour contains white pigments such as titanium dioxide (TiO_2), which become transparent like frosted glass when wet, so a colour appears. Although not a direct application of solvatochroism, this technology is now being intensively investigated by many textile companies.

3.6 Sleeping comfortably with sweet scents

A healthy and clean life has become a much more desirable objective in recent years. Department stores in Japan now provide an incense corner. Also many families in Japan perform a ceremony of burning incense before going to bed. People in Europe enjoy herb tea such as camomile in the evening to achieve a good sleep. Although the mechanism of sleeping is not yet completely elucidated, it is at least certain that sleep is closely related to brain activity, as indicated by electro-encephalogram readings. Scent is perceived by the cerebrum through an olfactory nerve, so that scent may also affect sleep.

According to the National Health Survey compiled by the Minister of Health and Welfare of Japan, mental stress leads to many health disorders. The survey recommends that a sound sleep is a good remedy for these symptoms and supports the maintenance of good health. Does not everybody wish to sleep comfortably and wake up pleasantly? The Institute of Sleeping Science, Japan, undertakes projects such as "the development of comfortable beddings", "the pleasant bedroom environment" and "the systematisation of the comfortable bed and bedroom environment". Various beddings have been investigated to promote physical and mental health. For example, some beddings have built-in magnets, ion generators; beddings with far infra-red radiation, etc. have been developed to maintain physical health.

The mental tranquillising effect of sound or scent is also being investigated to induce comfortable sleep, in order to promote better mental health. Some types of scent can induce good sleeping, and also unnecessary activity of the cerebrum can be eliminated by removing unpleasant odours. Thus a healthy life can be assisted partly at least by the use of scents for odorising and deodorising. Examples of deodorising methods are given in Chapter 4. Here we describe the scented fiber Cripy 65, developed by Mitsubishi Rayon.

The life of a scent essence is short, and heat accelerates scent loss. Thus for practical use the essence is captured in four petal-like cavities arranged radially in a hollowed fiber, produced by applying composite-spinning technology (see Fig. 3.14). The scent is gradually released through the fiber cross-section and the polymer film from the inside of the fiber and so

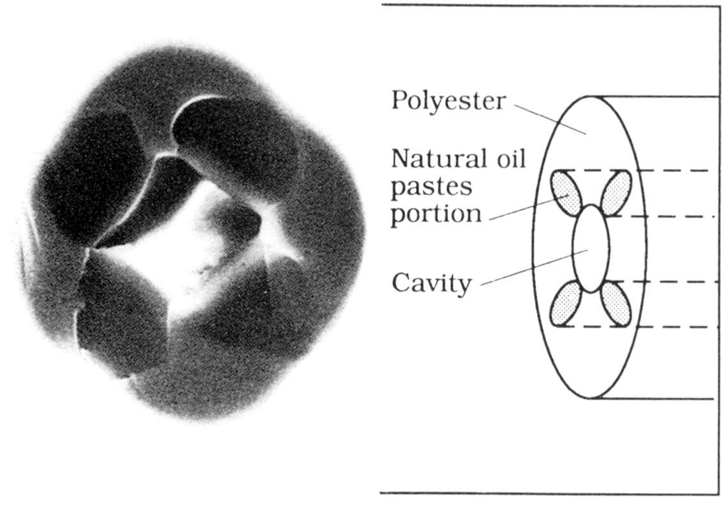

3.14 Cross-sectional view of Cripy 65 (Mitsubishi Rayon Co.).

maintains its effect for long periods. The scent essence of Cripy 65 was developed by joint research between Mitsubishi Rayon and Takasago Scent Ind., and is made up of more than 50 natural essences such as α-pinene, β-pinene and cedrol extracted from coniferous and lavender-type oil. The essence produces the refreshing effect typical of deep forests, and induces a good sleep, as shown by joint research between Toho University (Physiology Laboratory, Faculty of Medicine) and Tokyo Metropolitan Institute for Neurological Research.

Figure 3.15 shows the sleep stage diagram obtained by analysis of the type and depth of sleep as a function of time as measured by brain-wave patterns. Here the sleep depth is expressed in terms of four stages of non-REM sleep (designated by 1, 2, 3 and 4) and REM sleep (designated by R). The REM sleep and the stage 4 non-REM sleep increase when Cripy 65 is used for beddings. Although no quantitative method has been established to evaluate scent longevity in daily life, the scent of Cripy 65 lasts sufficiently long in the normal use according to users' tests. Such tests also confirm the positive effect of scent on satisfactory sleep.

3.7 Perfumed fibers

Perfume is widely associated with our daily life and is often used in religious ceremonies. Now more interest is being given to perfumes in the quest for a back-to-nature or healthy life.

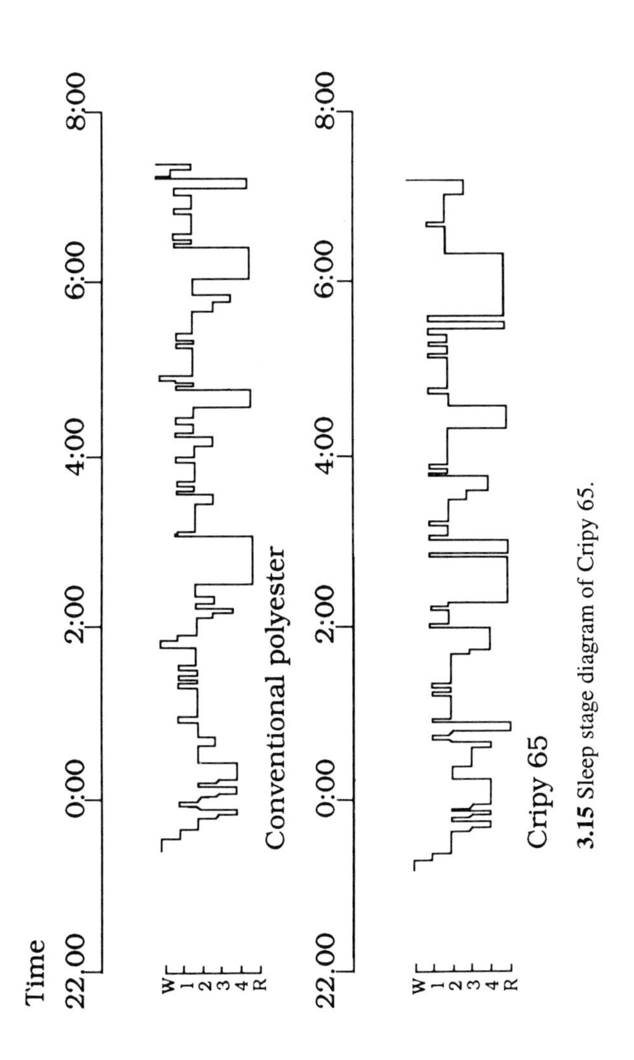

3.15 Sleep stage diagram of Cripy 65.

The perfumed fiber Esprit de Fleurs produced by Kanebo was first demonstrated at the "Exhibition of New Materials in Daily Life" held in Isetan Department Store, Tokyo, in 1987. The market of this perfumed fiber has expanded since then, and now sales are more than US$20,000,000 annually. Esprit de Fleurs is made of fibers to which resin-made microcapsules 5–10 μm in diameter containing perfume essence are bound. When the microcapsules are pressed and broken, the perfume is released. Microcapsule materials have been specifically developed such that the capsule thickness is accurately adjusted so that it does not break during fiber processing, but will break by friction during wearing. The material that binds these capsules to the textile must also be selected carefully so that the touch is not adversely affected (see Fig. 3.16).

Printing, dipping or padding is used for binding microcapsules to textiles, and this binding process is set at almost the last process in manufacture to prevent unnecessary loss of microcapsules due to breakage or falling off. The perfume lasts for at least two years, and can withstand several washings by hand or machine or dry cleaning. Applying the science of the aromacology, various kinds of perfume such as jasmin, rose, lavender, lily of the valley, sweetpea, greenflora, Indian sandalwood and a citrus flora can be provided for Esprit de Fleurs. Kanebo intends to develop a variety of applications, having found that perfume is also a major adjunct of fashion.

Esprit de Fleurs can be applied to various products including pullovers,

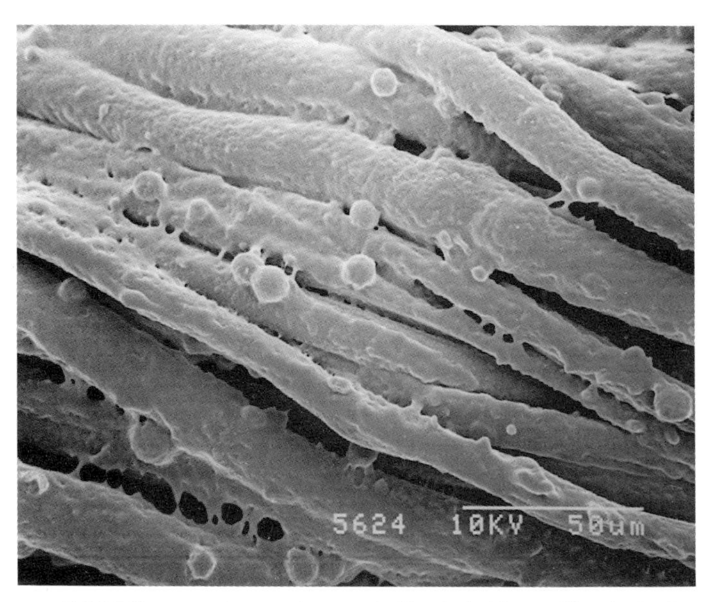

3.16 Microcapsules can be seen in Esprit de Fleurs (Kanebo Ltd).

cardigans, neckties, scarves, T-shirts, handkerchiefs, hand-knitting wools, stockings, spectacle cleaners, kimonos and polyester clothes for industrial use.

3.8 Power fibers that store solar energy

Heat-regenerating fibers are produced from ceramic composites by applying heat insulation processing technology which utilises the far infra-red radiation effect of ceramics. When heated, ceramics radiate far infra-red radiation which penetrates into the material and heats it homogeneously by activating molecular motion. Zirconium, magnesium oxide or iron oxide can be blended into synthetic fibers, because these materials radiate *ca.* 60 mW far-infrared of wavelengths 8–14 μm at a body temperature of 36 °C. These heat-regenerating fibers are used for sportswear, bed-sheets, bed-cover materials, etc. Further applications will be found in films to keep freshness, heat-insulating sheets for agricultural use, materials for greenhouses, etc.

There are two possible ways to insulate heat. One is by using passive insulating material which encloses the body heat, and the other is an active way that absorbs heat from the outside. For example, wool-based materials insulate heat with an inner air layer. Far infra-red reflection materials use coated aluminium. The far infra-red radiation materials radiating from ceramics activated by body heat, belong to the former category of insulating materials. The active insulating materials are exemplified by electrically heated materials, where the electrical energy of a battery is transformed into heat energy by the heater, with this heat supplied by the heat of oxidation of iron powder.

A futuristic fiber material Solar-α has been developed by Descente and Unitika, which absorbs and preserves the optical energy of the sun. Solar-α has been employed for a downhill skiing suit. The official uniforms of seven national teams were made from this material, first at the 1988 Winter Olympic Games and subsequently. The fiber has now proved its value world-wide. In addition to its smooth surface and aerodynamic form, this downhill suit aimed to increase the insulating efficiency by Solar-α in order to warm the muscle and bring out its best power. Solar-α was awarded the Technical Originality Prize for its novel insulating system at the 1988 Winter Goods Originality Competition held in Grenoble.

Oxygen consumption must be reduced to a minimum to bring out power efficiently from muscle in severe climatic conditions. In order to suppress the consumption of oxygen stored in the muscle, insulating efficiency plays a vital role in winter sports, in addition to ensuring water repellency and waterproof ability.

Zirconium carbide compounds are used for their excellent characteristics in absorbing and storing heat in a new type of solar system, including domestic water heaters and large-scale generators. Sunlight is absorbed in the heat condenser, and is converted into heat energy which drives a generator turbine. The maximum temperature obtained from conventional black-coated heat condensers is 90 °C, since the condenser loses part of the collected heat to the outside, owing to the temperature differential. Zirconium carbide traps heat energy. It absorbs visible rays and reflects infrared radiation. Zirconium carbide characteristically reflects the light of long wavelengths over 2 μm, but absorbs light energy of rather short wavelength (<2 μm), which makes up 95% of sunlight, and converts it into stored heat energy. Descente researchers applied these characteristics of zirconium carbide to achieve "active insulation", by enclosing microparticles of zirconium carbide in polyamide or polyester fibers. They developed the technology to enclose zirconium carbide powder within the core of synthetic fibers, in cooperation with Unitika (see Fig. 3.17). The clothes made of this fiber Solar-α absorb solar visible radiation efficiently and convert it into heat in the form of infra-red radiation which is released in the clothing. The released heat and the heat radiated from the body

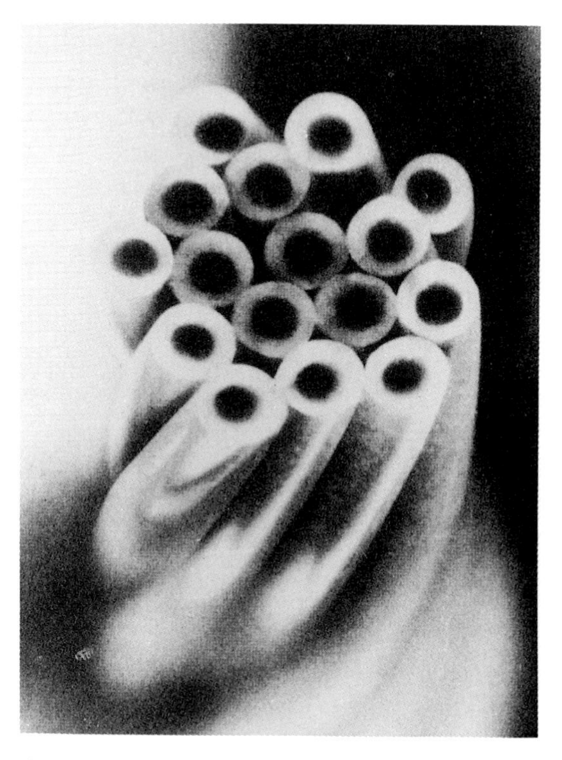

3.17 Cross-sectional view of Solar-α (Descente and Unitika).

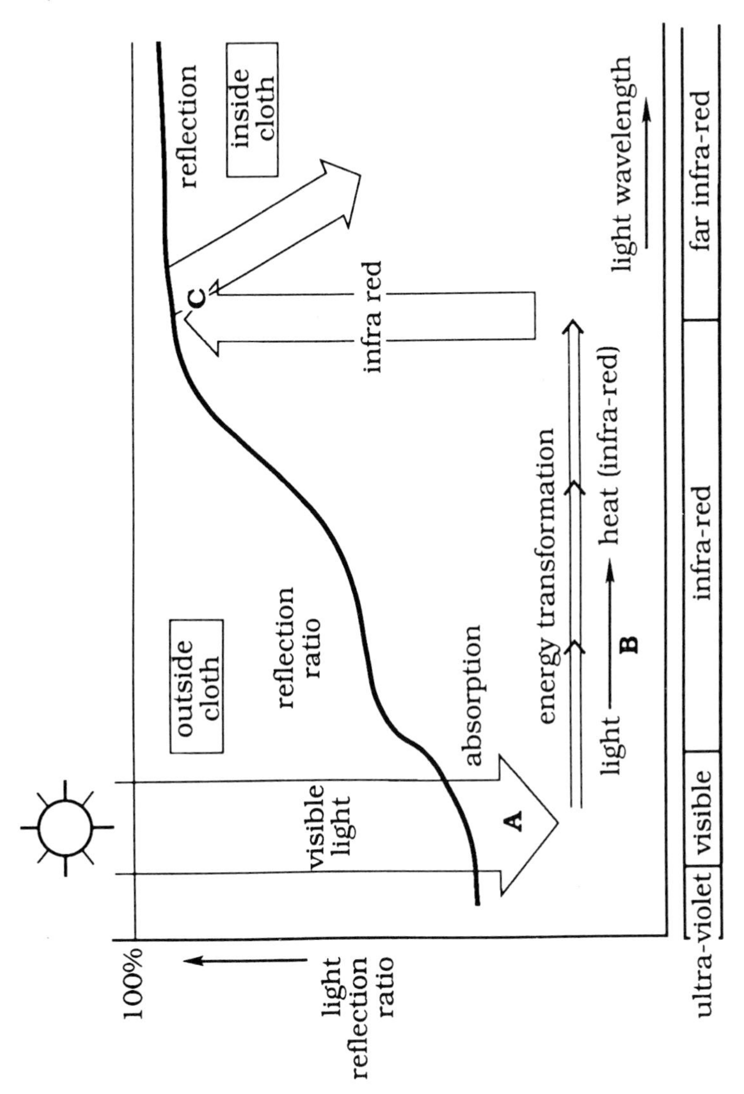

3.18 Heat absorption and insulation of Solar-α (Descente).

(far infra-red radiation) are reflected by Solar-α and will not escape into the outside (Fig. 3.18). In this way Solar-α sportswear suppresses the oxygen consumption in the muscle and brings out full power, even in extremely cold conditions.

3.9 Iridescent textiles

Nature has a great deal to teach us about the way in which colour can be exploited. Take, for example, a pearl shell, the peacock's feathers and the Brazilian *Morpho ala* which change in colour, according to the observation angle, owing to light interference. A pearl shell has a multi-layered structure whereas the peacock's feathers exhibit colour using a lattice-arranged system of fine melanin particles. *Morpho ala* reveals a metallic cobalt-blue colour owing to parallel ditches formed by its ladder-arranged scales.

Professor K. Matsumoto was attracted by such iridescence, and developed iridescent textiles by the use of light interference. Iridescent colouring caused by light interference can be observed using a polarising microscope. Professor Matsumoto investigated the colouring mechanism of drawn polymer films sandwiched between polarising films, and found a quantitative relation between the film thickness and the resultant hue. Fig. 3.19 shows the colouring mechanism of such a transmitting-type iridescent film. Here light is line-polarised through the first polyvinyl-alcohol polariser, rotated into eliptically polarised light through a 45° oriented anisotropic polymer film, and

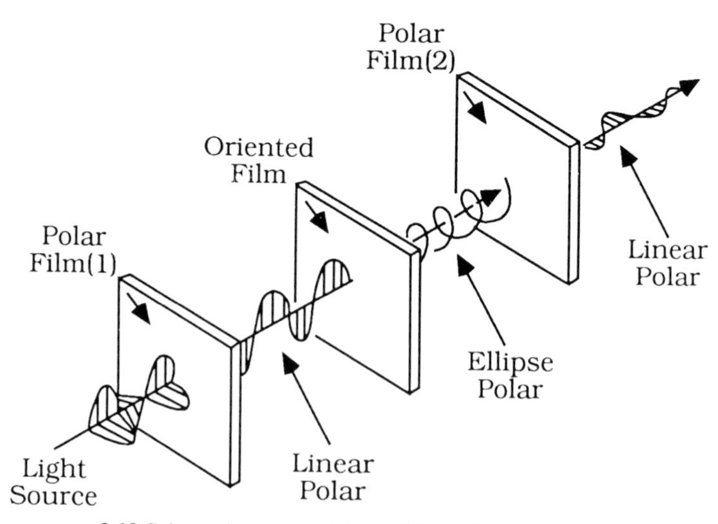

3.19 Schematic representation of the optical system of a transmitting-type iridescent film. (The arrows show the molecular orientation axes.).

then again line-polarised and coloured through the second polarising film. The delineation and elipticity of polarised light depends on the wavelength, and the transmitted light includes various wavelengths of arbitrary compositions which then exhibit the colour variations of the rainbow. This type of film is termed a transmitting-type iridescent film. When the second polarising film is replaced by a reflective film (for example, an aluminium leaf), the reflected light is also iridescent. This type of iridescent multi-layer film is referred to as a reflecting-type iridescent film. Figure 3.20 shows the colouring mechanism of such a reflecting-type iridescent film. These films are photo-controllable colouring films.

The transmitting- or reflecting-type iridescent film is sliced in fiber-like films 0.2–0.5 μm wide by using the technology used to make gold or silver threads. The fiber exhibits the hue required. Since the optical path-length changes with the visual angle or twisting, a single fiber can also be iridescent. Accordingly, a twisted yarn made of these fibers exhibits a variety of hues due the interference of the incident light. This iridescent yarn is expected to be applied widely in various textile fabrics (woven fabric, knitted fabric and lace, etc.). This is another development achieved by humans through learning from nature.

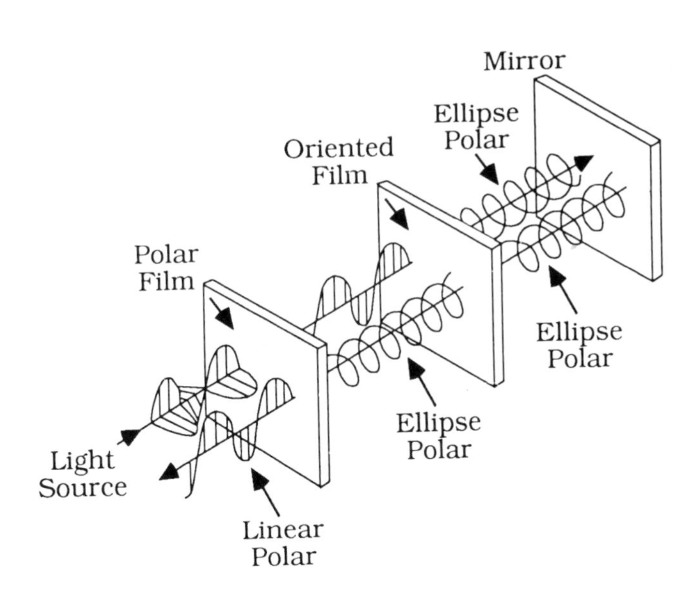

3.20 Schematic representation of the optical system of the reflecting-type iridescent film. (The arrows show the molecular orientation axes.).

3.10 Protein plastics with the feel of human skin

Natural leather is a complex assembly of collagen fiber bundles, which gives to leather its special texture, beauty, characteristic lustre and soft touch. A high-touch protein leather developed by Idemitsu Petrochemical Co. Ltd is made from pulverised collagen fibers that have been amalgamated with other plastics using an advanced compounding technology, as schematically illustrated in Fig. 3.21. This protein leather has the characteristic touch of natural leather, combined with outstanding toughness and durability. The material combines the merits of each composite component to give a warm, soft touch, good moisture absorption-release, but holds no static electricity like the natural product. It is water-resistant, easy to process and has the mechanical strength of a synthetic plastic.

Bovine skin provides the best available natural protein of consistent quality, and is made up of collagen (80 wt%), other water-soluble proteins and fat. The collagen has a molecular weight of approximately 300,000 and is made up of a rod-like triple helix 3000 Å long and 15 Å in diameter. These rod-like collagen helices align regularly to form non-soluble collagen fibers. Pulverisation of such collagen fibers to a size of a few micrometres has been attempted in the past, but without industrial success. However, the Idemitsu Petrochemical Co. developed a new two-step process to produce a fine homogeneous powder of collagen fibers. In the first stage the collagen fibers are processed to prevent their gelation, and the second stage prevents their fine fiberisation during pulverisation. The powder is produced in diameters varying from 5 to 100 μm in four grades, according to the desired application. These now extend from protein plastics to protein paints.

Polyurethanes, polyvinyl chlorides (PVCs) and polyolefins are the main matrix materials used for the protein plastics, which are moulded at low temperatures. This composite protein leather is further processed into products, such as shoes, building materials, clothing, sportswear, bags, interior decorations, etc. Four types of protein leather with different touch are commercially available according to the application.

One example is the urethane impregnation type Grancuir which has been applied in sporting goods, high-quality furniture, etc. Grancuir is made up of an artificial leather layer of unwoven ultrafine polyester fabric impregnated with porous polyurethane and collagen fine powder with an average diameter of 5 μm. This is laminated with a surface layer composed of polyurethane and collagen fine powder (see Fig. 3.22). The collagen represents about 0.1 by weight of the total leather product. The collagen is impregnated in this way even in the back (unwoven fabric) side of the leather, in order to retain moisture absorption-release functions. Since conventional artificial leather is

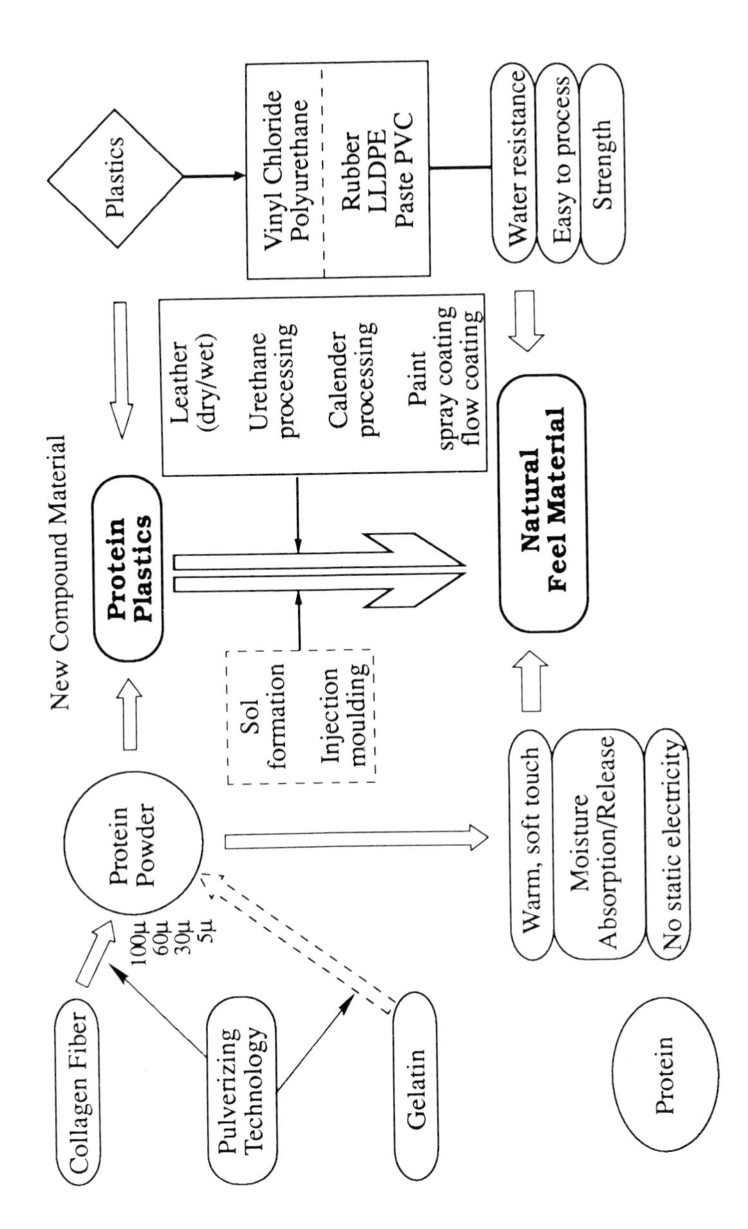

3.21 Collagen fibers in natural feel materials.

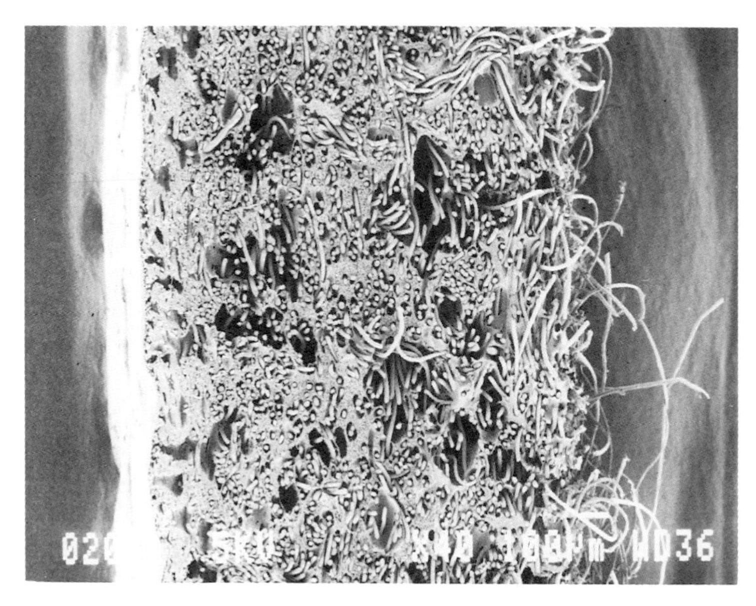

3.22 Cross-section of Grancuir (× 30 scanning electron microscope), and Grancuir shoes.

made up of hydrophobic synthetic fiber and polyurethane, its moisture permeability is low (2 mg/cm^2 h) compared with natural leather (13 mg/cm^2 h). Grancuir, however, possesses high moisture permeability (9.5 mg/cm^2 h) and good moisture absorption-release characteristics, so this material is resistant to wrinkling and feels clean and pleasant. Grancuir appears in all respects to be identical to real leather material, but is tougher, and has greater durability and better water-resistance. The good moisture permeability of Grancuir is attributable to the ready diffusion of moisture through the polyurethane pores and a pumping effect of collagen component which absorbs moisture from the high-temperature side (the inside of a shoe) and exhausts to the low-temperature side (the outside of the shoe). It therefore provides another good example of the fiber technologist being able to emulate nature and provide a material that is even more suitable for practical purposes than that directly fabricated from animal origin.

4 Biomimetic chemistry and fibers

The Nobel Prize for Chemistry in 1987 was jointly awarded to Dr Pedersen, formerly of Du Pont, Professor Cram of California University and Professor Lehn of Pasteur University, who pioneered biomimetic chemistry. Dr Pedersen synthesised a large-ring compound designated a "crown ether" (Fig. 4.1) in 1967, which was later theoretically investigated by Professors Cram and Lehn, and resulted in opening up a new field of chemistry. This research showed the potential for producing artificial enzymes and cell membranes. Subsequently, many investigations have been undertaken to mimic biological functions. As a result, not only single but also cooperative biological functions can be realised artificially.

The term "biomimetic chemistry" was proposed by Professor Breslow of Columbia University in 1972 to distinguish this new research field. Here "bio" denotes the living system, and "mimetic" its imitation. Professor Breslow first investigated the enzymic functions of crown ethers, and later developed the synthesis of macromolecules whose functions could be applied practically. Professor Breslow reproduced chemically enzymic reactions by using model systems, and Dr Pedersen synthesised materials that are

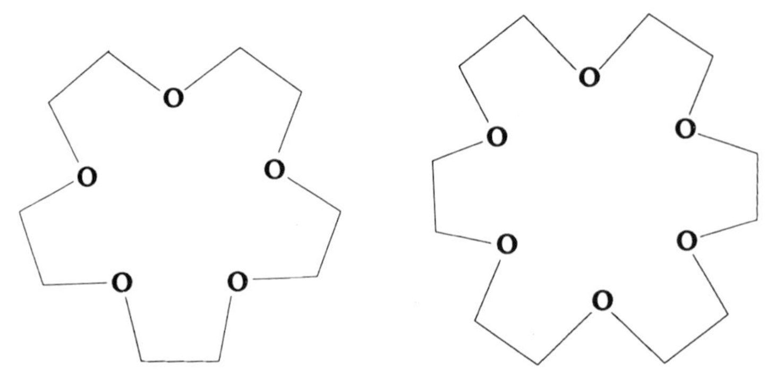

4.1 Large ring compound "crown ether".

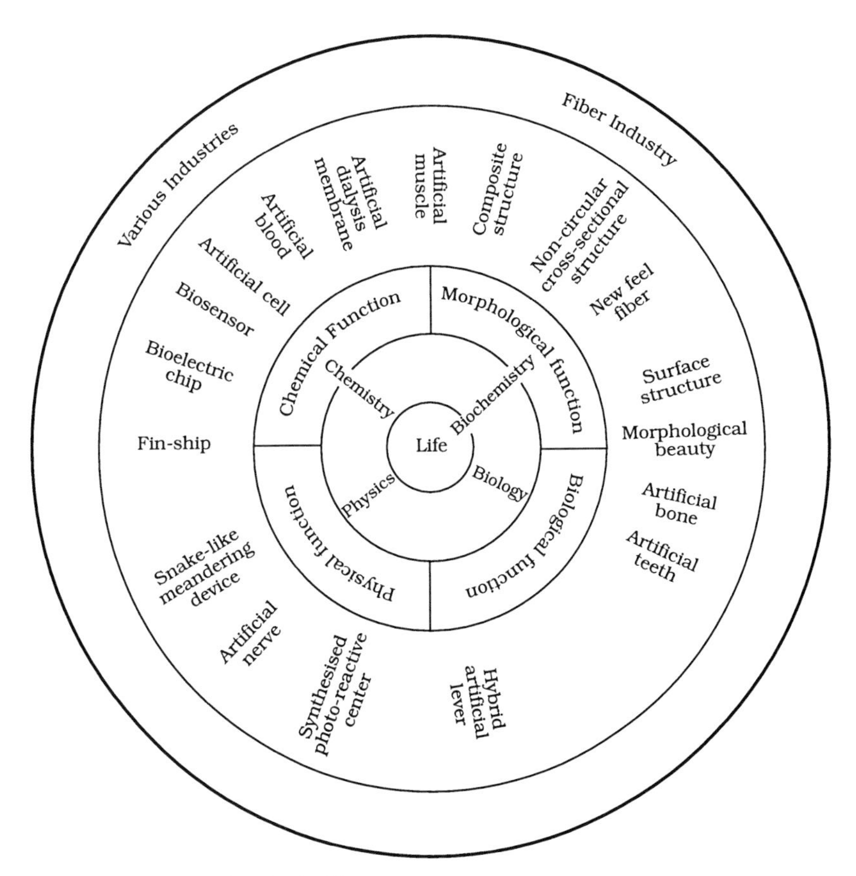

4.2 Biomimetics in fibers.

analogous with organisms in their ability to react selectively with particular ions or molecules. This field of research and technology is now termed "biomimetics". Since the living organism is too complex and subtle to imitate as a whole, many attempts have been made to reproduce the functions of a part of the living organism with synthesised material. "Functions" here constitute important characteristics of living organisms. Fibers in living systems relate to morphological, biological, chemical and physical functions, as indicated in Fig. 4.2. This chapter outlines the morphological and biological functions, as related to the fiber industry.

4.1 Applications of morphology/structure

The structure and functions of biological materials are precise and subtle. However, if the way in which structure and function are related can be identified, then it might be possible to replicate this molecular design in the service of the fiber industry. For this purpose the fiber industry may conveniently be divided into:

1 The spinning industry of mainly natural fibers such as, cotton, wool and silk.
2 The chemical man-made synthetic fiber industry.

There has been a positive interaction between these two areas. Pioneers, fascinated with silk, made great efforts to produce a silk-like fiber by dissolving cellulose and spinning its solution. The name "rayon" implies that it looked like a "ray". The patents applied for by industry relate the history and trend of the technical developments in this field. Many of the synthetic fibers were developed to imitate the structure and characteristics of natural fibers. For example, a fiber with a silky-lustre and smooth hand-feel was produced by copying the triangular cross-section of silk. A fiber with a deformed cross-section was developed to prepare a cotton-like macaroni-type fiber. This technology was applied to the industrial production of separation films and dialysis membranes for artificial kidneys, which is one of the greatest achievements of present-day high technology.

When comparing the patents applied for by Du Pont with those of Japanese companies, it is evident that Du Pont have placed their emphasis on fundamental research and have been quick to learn from nature. Their Japanese counterparts have been quick to export these ideas to develop the field in a different direction. For example, the silk-like fiber with deformed cross-section produced in Japan has its roots in "Quiana", developed by Du Pont. Many industries in Japan developed silk-like processing technology by combining the spinning of fibers with triangular cross-section and their blending with fibers having a different crimp ratio.

This spinning and processing technology for two components with different characteristics was developed by copying the double layer structure of wool. An extension of these ideas led to the sea–island composite spinning technology described in Chapter 3. This has further been utilised to produce optical and antistatic fibers.

Composite spinning was also applied to produce artificial leather. Natural leather is composed of layers of bundle assemblies, with finer fibers approaching the surface from the inner skin side. At the surface these fine fibers are disbundled and entangled. This structure was successfully imitated by deliberately entangling bundles of fine fibers. Thus, many new artificial leather products have now appeared (Fig. 4.3). Through first imitating natural fibers, natural fibers have now been surpassed. For example, the super-fibers such as carbon, heat-resistant, glass, ceramic and amorphous fiber are now better than the original. It has, therefore, become a biomimetic industry.

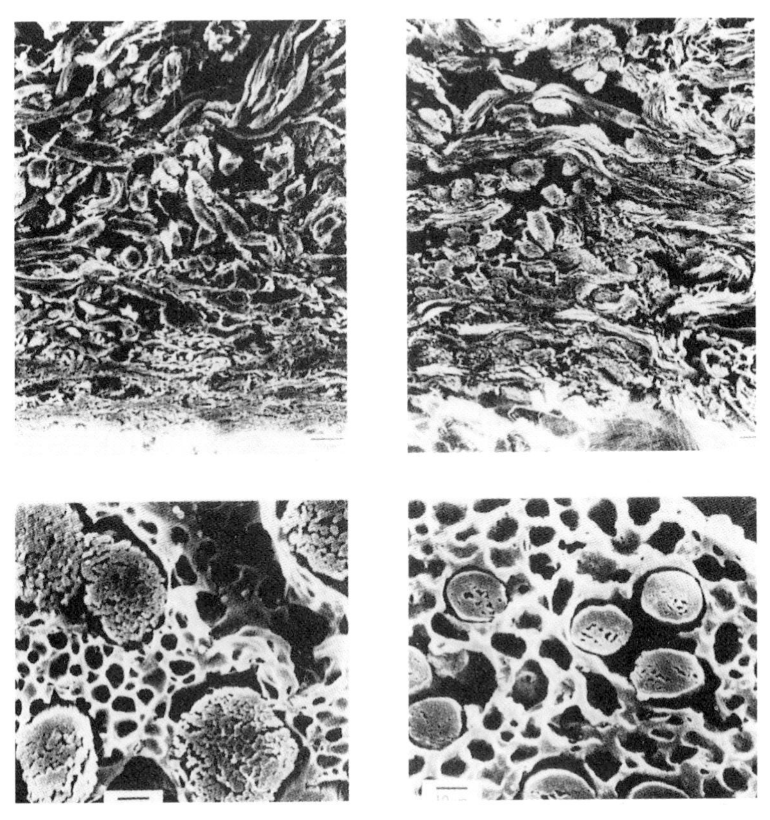

4.3 Which is genuine? The top left and top right show the sectional structure of suede-type leather. *Top left:* natural leather. *Top right:* artificial leather Kuraline F (Kuraray Co.). *Bottom left:* full-grain soft (artificial) leather Kuraline F (Kuraray Co.). *Bottom right:* hard (artificial) leather Kuraline for shoes (Kuraray Co.).

Fiber science and fiber technology can be related to biomimetics, as indicated in Fig. 4.4, where the overlapping part A denotes specific technologies based on three different fields of knowledge. In this way new technologies concerning conjugated fibers, scrooping, water repellency and deep colouring effects were developed. Table 4.1 shows examples of biomimetics in the fiber industry which yielded new products by applying the morphological functions of living organisms. This frontier of fiber technology will now be reviewed.

4.1.1 Composite structure of natural fur

Toray has developed a mink-like artificial fur Furtastic which imitates the composite structure of natural fur, and follows on from its suede-type material

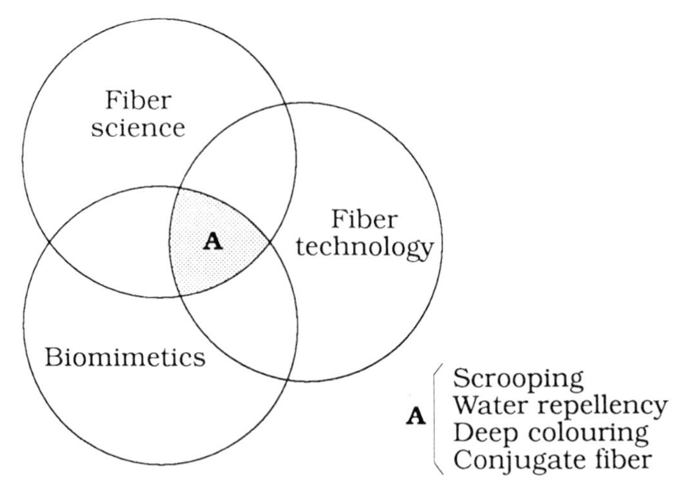

4.4 Interaction of fiber science, fiber technology and biomimetics.

Table 4.1. Examples of biomimetics in fiber industries

Function	Structure	Living system	Biomimetics
	Composite structure	Natural leather	Ecsaine (Toray Industry Inc.)
Morphology	Non-circular cross-sectional structure	Silk, wool	Grasem (Kanebo Ltd.)
	Surface structure	Lotus leaf	Microft-Lectus (Teijin Co.)
	Morphological beauty	Butterfly, moth	Dephorl (Kuraray Co.) Microcrater (Kuraray Co.)

Ecsaine. Observed microscopically, it is seen to consist of piles of extremely fine fibers and "stinging hairs", densely assembled and standing on a skin as in natural mink fur. Furtastic has a double-layer structure composed of "stinging hairs" and "piles" of sharp ends which are densely planted on the skin base as shown in Fig. 4.5 and 4.6.

The know-how developed for Ecsaine production was applied to prepare the skin base and the piles of extremely fine fibers. The "stinging hairs" were produced using a newly established technology developed specifically for this purpose. A second technique developed specially for Furtastic is "flocking", to prepare a textile with high raising density. Toray developed flocking technology that could increase the raising density to 17,000 flocks per square centimetre. A patch of natural fur from a single mink is *ca.* 10 cm wide, and the back skins of these patches are stitched together to make the fur broader and bulkier into the coat. Furtastic is produced as a broad sheet, but in a way which looks as though smaller patches had been stitched together. This can be achieved by changing colours carefully. This takes account of the fact that each mink has its specific colour characteristics. Toray has in this way

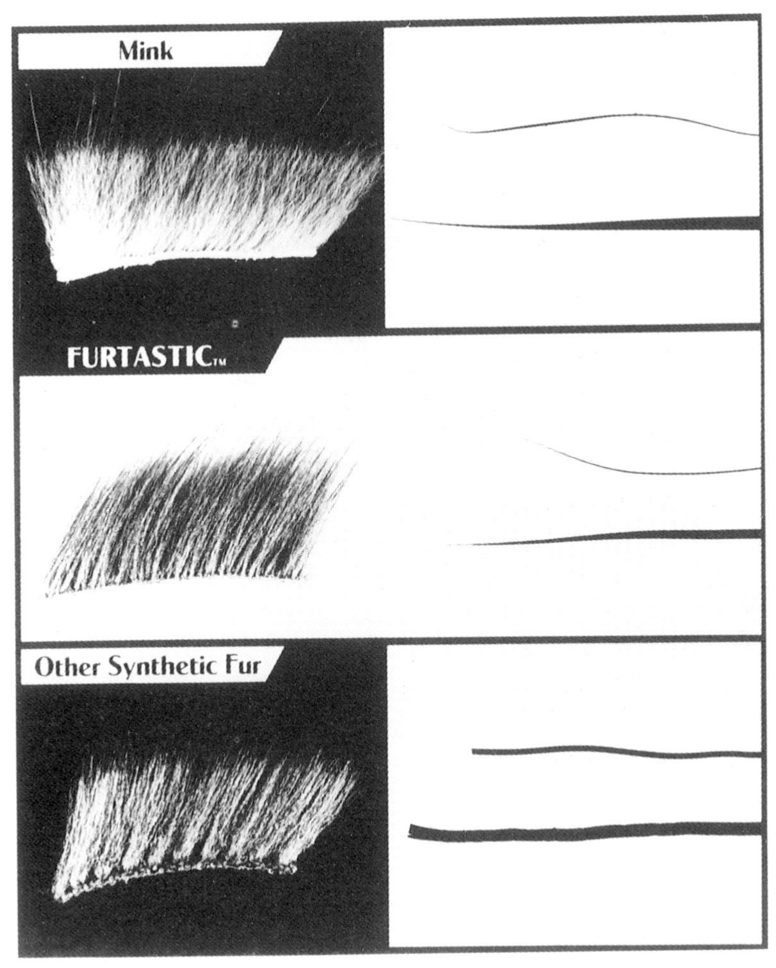

4.5 Side view of the standing hairs and hair-ends in Furtastic (Toray Industry Co.).

produced more than 36 000 m^2 of artificial fur, which has mainly been employed for sports- and ladies-wear.

Most of the full-grain artificial leathers are produced by nylon-coating PVC leather, or urethane-coating non-woven fabrics. These products are mainly used for making bags or shoes. However, the moisture vapour transmission and hand feel is not as good as full-grain natural leather, because of the polymer-coated surface. Natural leathers can be divided into either (i) the raised suede-type or (ii) the full grain sheep-type. The sheep-type full-grain artificial leather Youest was produced to compensate for the defects in conventional artificial leathers. Ecsaine and Furtastic belong to type (i) and Youest as noted to type (ii). Toray made up the surface of Youest from

4.6 Three-layered fabric Furtastic (Toray Industry Co.).

high-density fabric made up of extremely fine fibers. When fabric is made from such very fine fibers, it looks like a film to the naked eye but its structure closely resembles that of natural sheep leather (see Fig. 4.7).

Extremely fine filament (0.001 denier) is used for Youest, whereas less fine filament (0.1–0.05 denier) is used for Ecsaine. Youest is thin enough to be used for clothing materials having a soft skin-like feeling.

Toray also developed the micro-pocket fabric Toraysee by applying a new microfiberisation technology. Ecsaine, Furtastic and Youest are non-woven fabrics, but Toraysee is a high-density woven fabric with ultra-fine fibers, which possesses pores containing micro-pockets. Since dust or oil is absorbed by these micro-pockets, Toraysee is well suited for cleaning optical lenses as described in Chapter 3. Some 700 ultra-fine polyester filaments ($\phi = 0.01$ mm) are bundled and woven to make Toraysee without causing fuzzing.

Toray also developed the electrically charged non-woven fabric Toraymicron with a highly oriented polarised structure. Dust or bacteria is charged either positive or negative. Thus these charged micro-particles are caught electrically as well as physically as with conventional filters of non-woven fabrics, when filtered through Toraymicron. This material is made from non-woven fabrics of ultra-fine fibers, using the same techniques as for

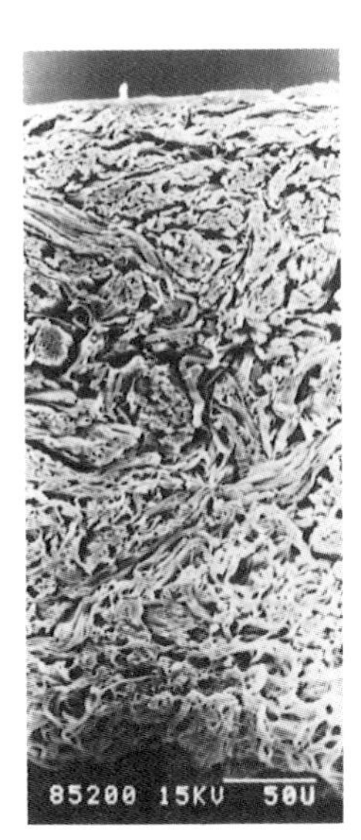

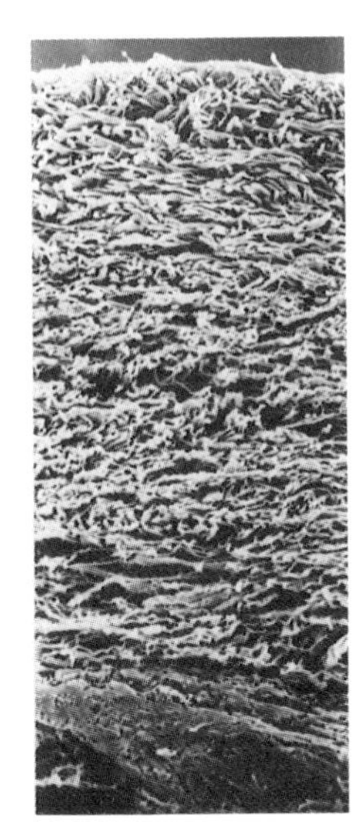

4.7 Cross-sectional view of Youest: *left*, Youest; *centre*, natural leather; *right*, conventional artificial leather (Toray Industry Co.).

Ecsaine. A better filtration efficiency would be expected with a finer-mesh assembly of thin filaments. The filtration efficiency of Toraymicron is further improved by its electrical properties which increase the dust-collecting ability without causing pressure loss. Toraymicron is, therefore, used for the highly efficient filters for clean rooms, masks, machine filters and filters for domestic vacuum cleaners.

4.1.2 Super-Microft based on the structure of a lotus leaf

In 1987 the Technology Prize of the Society of Fiber Science and Technology, Japan, was awarded to the Teijin Co. for its development of fabrics with high water repellency. Super-Microft is one that was designed by emulating the structure of a lotus leaf. Water rolls like mercury from the lotus leaf, whose surface is microscopically rough and covered with a wax-like substance with low surface tension. When water is dropped on to the surface of a lotus leaf; air is trapped in the dents and forms a boundary with water. The contact angle of the water is large, because of the wax-like substance (see Fig. 4.8 and 4.9). The apparent contact angle depends on the evenness and roughness of the surface, and surface tension. When (i) the surface is reasonably even but with

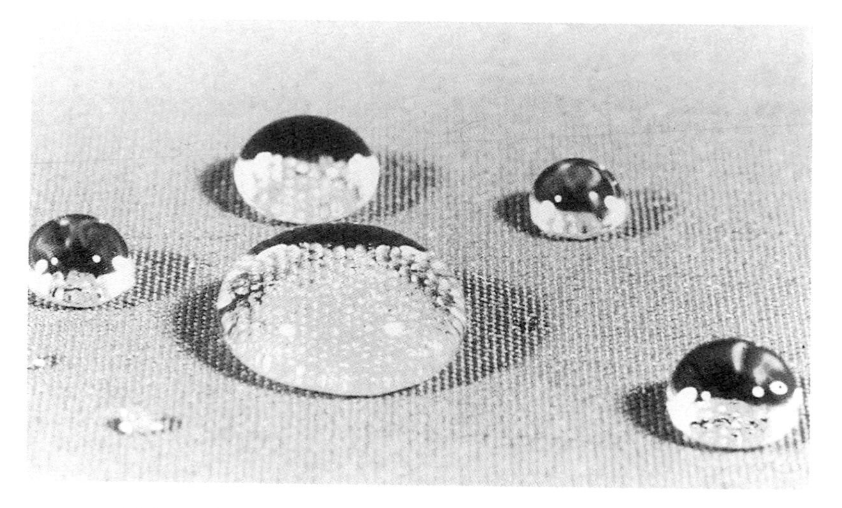

4.8 Super-Microft on which water rolls like mercury (Teijin Ltd).

microscopic dents to enlarge the surface area and to trap air, and (ii) the surface tension of fibers is small, then water rolls well on the fabrics. Super-Microft is a highly water-repellent fabric made of polyester fibers, harnessing the water-repellent mechanism of lotus leaves.

The technology of Super-Microft production consists of:

1 The design of the original filaments (to allow air-trapping with high durability, partial crimping, and the potential to give a high bulky filament).
2 The textile design (to give a high density textile with a natural cotton hand-feel and good size stability).
3 A new dyeing process (dyeing in such a way as to produce homogeneous microscopic dents on the textile surface).
4 A new finishing process to reduce the surface tension (by combining water-repellent and wash–wear-resistant processing).

Super-Microft exhibits good water-repellent durability and a high wear resistance. The criterion required for water repellency is to have a "rolling angle" which is extremely small (less than 10°) even after five days of washing in comparison with conventional fabrics (30–40°). This ensures that a water drop rolls well on Super-Microft.

Water-repellent fabrics are widely applied in the production of outdoor sportswear (windbreaker, ski-wear), and general clothing (coat, working clothes) as well as industrial fabrics (tent, bags). So far, only water repellency that relates to the waterproof ability of fabrics has been considered. However, water repellency can also be utilised to evaporate moisture coming from the human body. The new water-repellent fabrics described above are both

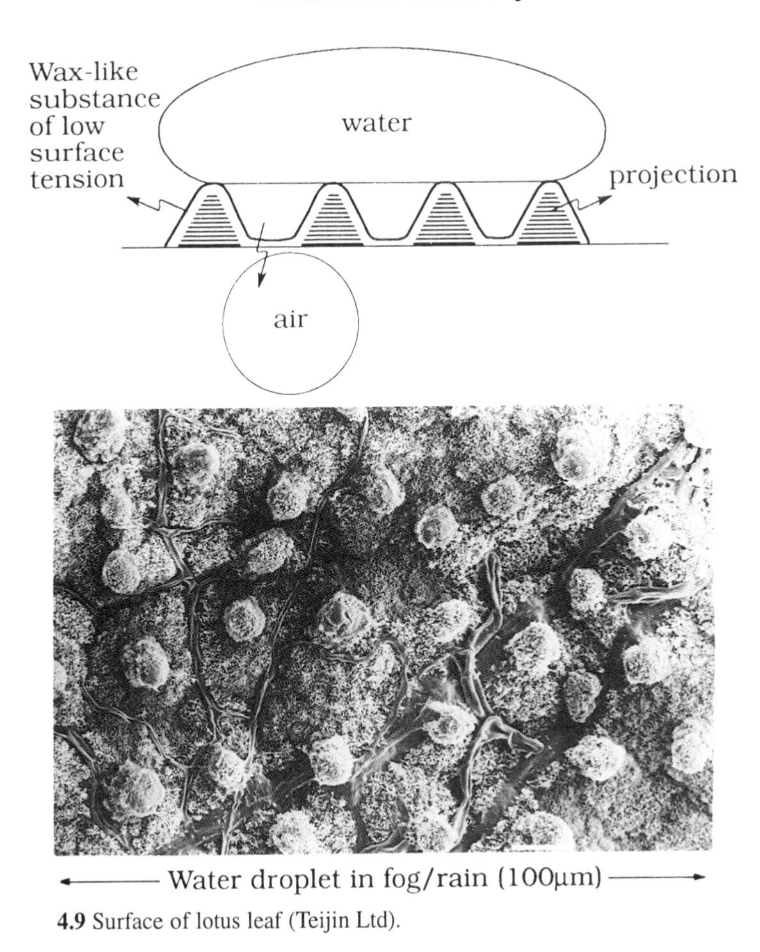

4.9 Surface of lotus leaf (Teijin Ltd).

moisture-permeable and waterproof. Therefore, ski-wear from this material does not become stuffy, as does a conventional raincoat. To achieve this, water should form droplets on the surface and not spread evenly over clothes, and thus prevent a water layer forming which could stop moisture permeation. Water repellency must be considered within this overall context.

4.1.3 Morpho-structured fabrics imitate the insect morpho alae

Kuraray developed a new textile Morpho-Structured Fabrics by closely observing and reproducing the structure of the insect *Morpho aloe*. Many insects exhibit various colours according to their natural environment. The colouring mechanism control is so clever that it can teach us more about how to develop materials for clothing. Morphos (Fig. 4.10), which inhabit the

4.10 Morpho ala.

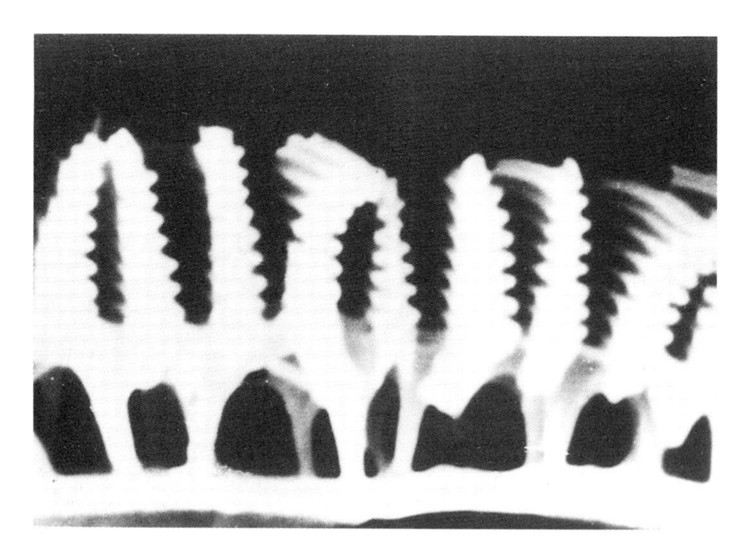

4.11 Scale structure of *morpho alae*.

Amazon valley, are one of the most beautiful genus of butterflies, with a metallic cobalt-blue colour. *Morpho hecuba* is the largest of the species, about 17 cm long. The surface structure of a butterfly alae indicates the fundamental role which the reflection ratio plays in colour appearance as a result of light interference. Since ordinary fibers have a low reflection ratio, they fail to show brilliant colours even when dyed. The colour of *Morpho alae* changes subtly according to the angle of incident light. Its scales are both coloured and non-coloured. The coloured scales are *ca.* 2 μm in height, with flat wall-like appendages with nine or ten pleats running in parallel at regular intervals of *ca.* 0.7 μm as shown in Fig. 4.11. The incident light will reflect/refract/interfere at these appendages, and brilliant metallic colours appear.

Morpho-Structured Fabrics were developed by copying the structure of *Morpho alae*. When bicomponents of different thermal properties are spun, shrinkage will occur, and the resultant fiber is effectively twisted. Morpho-Structured Fabrics are made from a twisted fiber of flat cross-section, with 80–120 twists per inch, which is achieved by thermal processing. When woven and then thermally treated, fibers are twisted and the flat surfaces of the fibers align vertically by reference to the woven fabric plane (see Fig. 4.12). The deep-colouring of Morpho-Structured Fabrics are due to the alternative horizontal/vertical alignment of the flat surfaces of fibers which cause the repeated reflection/absorption of the incident light and reduce the direct reflection (see Fig. 4.13). Thus Morpho-Structured Fabrics exhibit a deep brilliant colour, and also have soft and elegant drape characteristics, suited to dresses and blouses, which imparts a new feeling. The deep slim ditches in the surface of "Diphorl" give rise to a full variety of deep and brilliant colours that conventional fabrics are unable to achieve.

4.12 Morpho-Structured Fabrics: potentially twisted fibers of flat cross-section (Kuraray Co.).

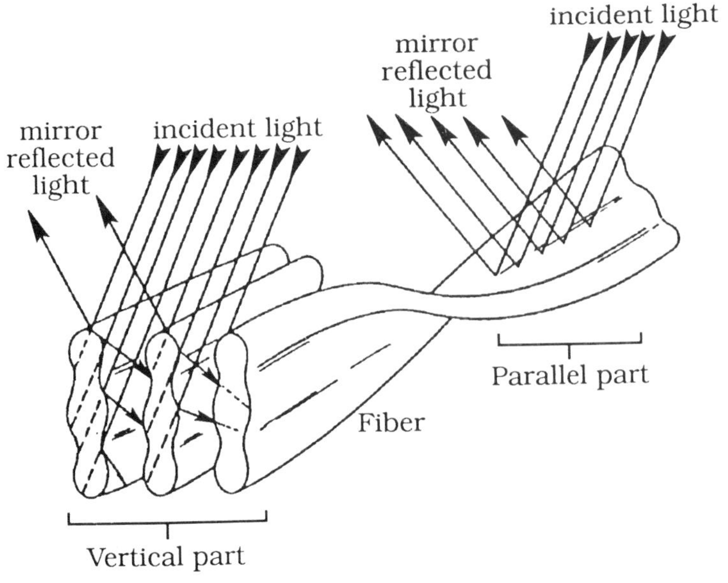

4.13 Deep colouring effect of Diphorl.

4.1.4 Super-Microcrater Fiber: imitating the cornea of the night-moth

Kuraray developed Super-Microcrater Fiber which can achieve a deeper colour effect by emulating the cornea of a nocturnal insect (night-moth). Nocturnal insects have an astonishing natural mechanism that enables them to survive in the dark. Whereas a bat flies freely to catch insects in the dark using its ultrasonic sensor, a night-moth is equipped with a sensor to detect the ultrasonic radiation transmitted from a bat so that it can escape quickly. The cornea of a night-moth when closely observed, has a number of conical projections arranged hexagonally, whereas diurnal insects such as dragonflies and grasshoppers have no such projections. When a billion per cm^2 microcraters are constructed on the fiber surface like the cornea of a night-moth, these craters trap the incident light and consequently the fiber exhibits a deeper colour (see Fig. 4.14). Since the direct reflection of the light at the fiber surface is suppressed, the fiber is able to exhibit brilliant black which conventional fibers cannot achieve. Kuraray has greatly expanded its share of black formal-wear for ladies using Super-Microcrater Fiber.

A highly civilised society promotes individuality in its human life-style by variations in the use of clothing. This requires a variety of colours and forms according to individual sensitivity. We should learn more from the secrets of nature to satisfy the depths of diverse human consciousness.

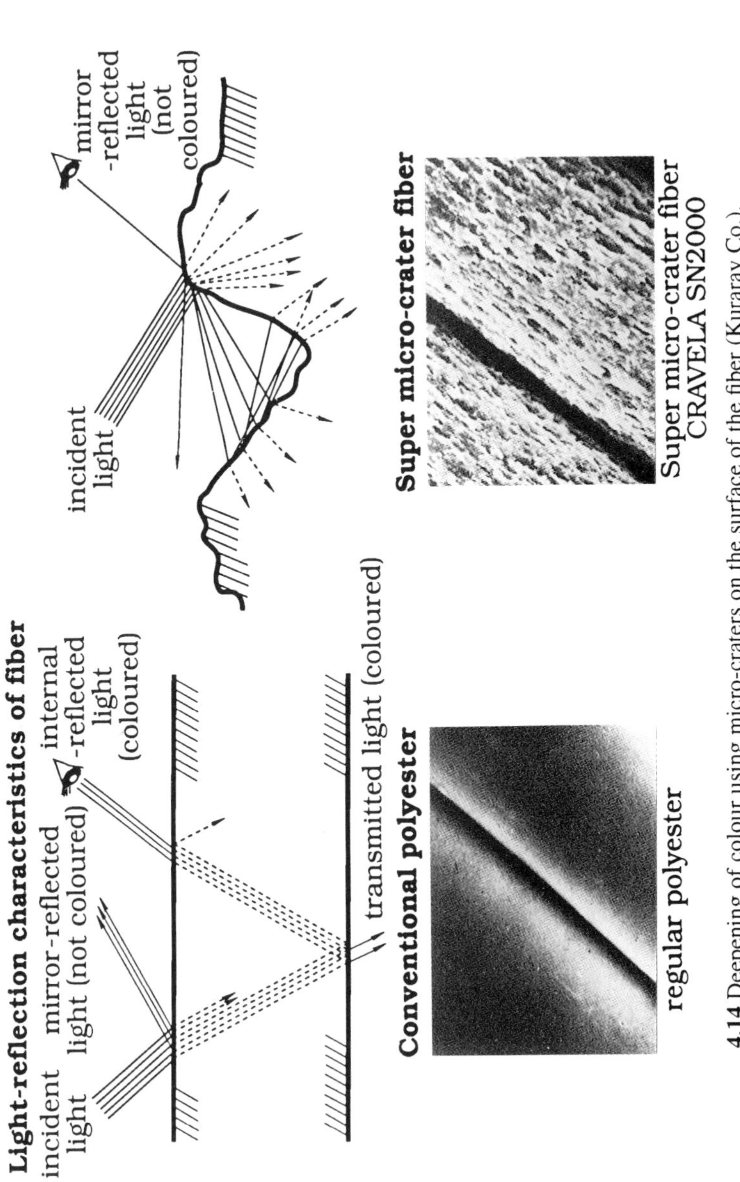

4.14 Deepening of colour using micro-craters on the surface of the fiber (Kuraray Co.).

4.1.5 Enzyme-like deodorant fibers

Proteins from meat or soya beans can be decomposed only when cooked with strong acid at high temperatures for long periods, but in the human body, enzymes work as catalysts to decompose proteins into amino acids effectively within a few hours at room temperature. Polymers and other chemicals used in everyday life are produced at high temperatures and pressures in chemical factories, which inevitably cause pollution. Enzymes are capable of producing chemical materials at ambient temperature and pressure without causing any pollution. Although enzymes are highly selective/energy conservative/non-pollution causing catalysts of high efficiency, as materials they are weak and relatively unstable to high temperatures, can be deactivated when contaminated with various bacteria, cannot be used in organic solvents and can be applied only to specific reactions. The development of artificial enzymes will depend on the ability to cope with their vulnerable characteristics. Various artificial enzymes have now been developed and used in everyday life. For example, emulating hematin (ferri (Fe^{3+}) protoporphyrin IX), Professor Shirai at Shinshu University developed a deodorant fiber that has 100 times better efficiency than active carbon. In collaboration with Daiwabo Co., it was harnessed into a commercial product, a deodorant bed "Green Life" which utilises deodorant fiber wool as filling.

Toxic substances taken into a human body are oxidised and detoxified by oxygen in the blood, owing to the catalytic function of a series of oxidising enzymes, including the haem protein. The haem protein contains a reactive centre, which is a flat ring-like ferri-compound called protoporphyrin. The enzyme structure is shown schematically in Fig. 4.15, and consists of a reactive centre, surrounded by the U-shaped part that provides the reactive environment, and supported by the moiety specifying the active site. This U-shaped protein of high molecular weight, which functions with great specificity, can accept or remove electrons, controls the reaction speed within the reactive-environment part, and governs the chemical reactions which occur at the reactive centre. This functional cooperation results in the following:

1 High activity.
2 Energy conservation.
3 High selectivity.
4 High efficiency of the enzyme.

The enzyme loses activity if a structural change accompanies any significant environment change. The enzymes thus function precisely and cleverly within the living body, where hundreds of chemical reactions proceed simultaneously

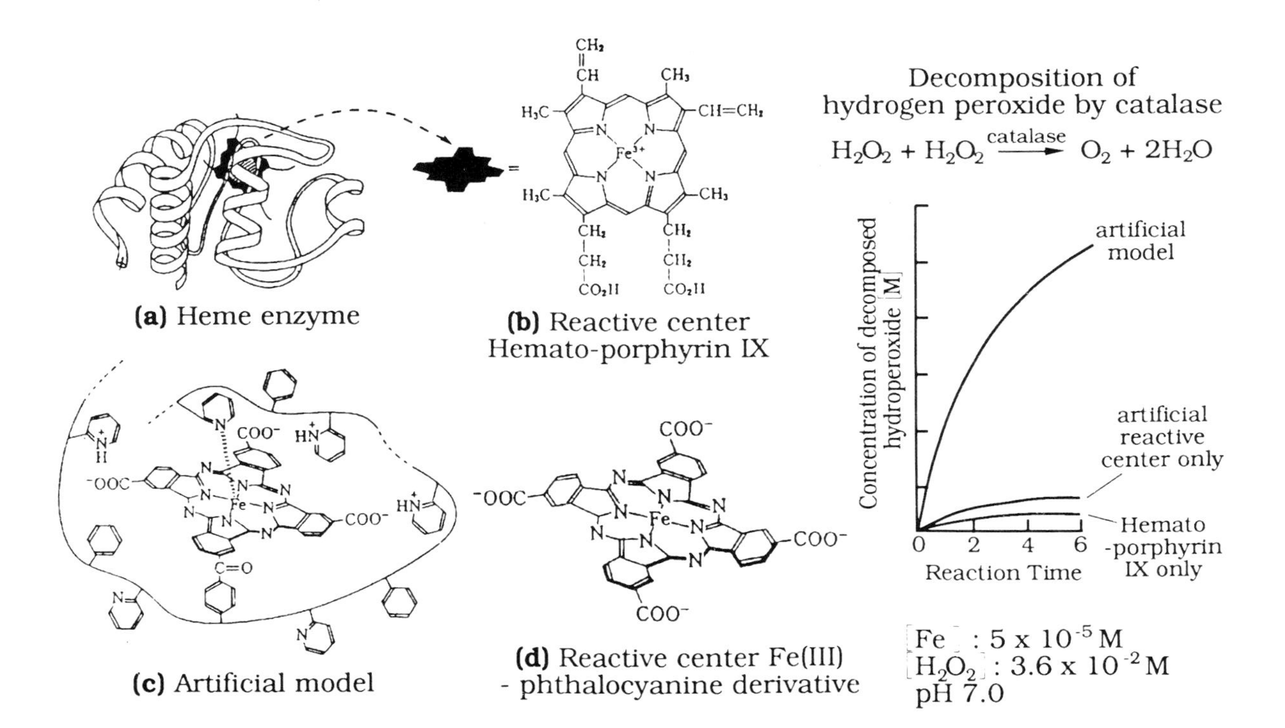

4.15 The performance of an artificial enzyme with an Fe(III)-phthalocyanine derivative reactive centre.

at normal temperature. Animals, including humans could not survive without such enzymes.

Professor Shirai imitated the haem enzyme using polystyrene derivatives coupled with Fe-phthalocyanine. This artificial enzyme has an Fe-phthalocyanine complex as its reactive centre instead of protoporphyrin IX, and although structurally different from the haem enzyme, exhibits a relatively high (1/50th of the enzyme) activity (see Fig. 4.15). In Japan phthalocyanine is widely known as the blue/green paint which forms a complex with copper, and is used for the body colour of the Shinkansen bullet trains.

The reactive centres must be fully exposed and dispersed homogeneously to make efficient contact with reactants to achieve efficient catalytic reactions. For this reason, amorphous rayon was chosen as the supporting material, because this rayon is porous (i.e. it possesses a large surface area) and is amphiphilic. The degree of crystallinity of amorphous rayon (15%) is smaller than conventional rayon (*ca.* 40%). The deodorant fiber consists of amorphous rayon supporting 3 wt% Fe-phthalocyanine type artificial enzyme. This deodorant fiber is capable of destroying an offensive odour by decomposing the foul smelling molecules, such as indole (faeces), hydrogen sulphide (rotten eggs) and mercaptans. The deodorant fiber maintains its activity about 100 times longer than active carbon, but is effective only with a limited range of malodorant sources as shown in Fig. 4.16. The deodorant fiber is used at present in commercial products such as nappies (diapers), an inner sole of a shoe, in refrigerators, and the mat for a toilet seat. Materials other than amorphous rayon are also being examined as a supporting material for artificial enzymes.

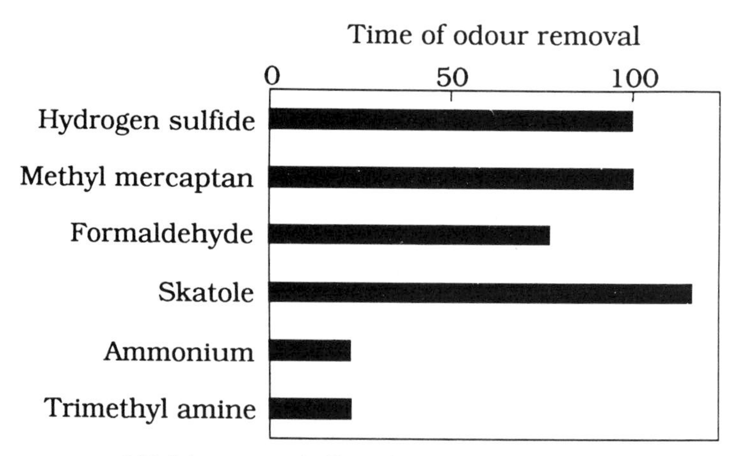

4.16 Odour-removal effect of Fe-phthalocyanine-doubled fiber with respect to that of active carbon (which is normalised to 1).

4.2 Hybridisation technology

The term hybrid" stems from the Latin "*hybrida*", which denotes the offspring of a tame sow and a wild boar, i.e. a mongrel. Many attempts have been made to hybridise synthetic polymers and biopolymers of a particular structure and function to produce a new material suitable for particular applications (see Fig. 4.17). This hybridisation is a common technology in the biomimetic industries as shown by the following examples.

4.2.1 Hybrid-type artificial organs

If a chemical plant were to replace the function of the liver, the plant would need to be as large as the Empire State Building. The liver is a finely controlled chemistry factory, and along with the heart, must be regarded as the most important organ in the body. The organ produces bile for digestion, stores glycogen for energy, decomposes toxic substances taken into the body, and controls the metabolism of lipids. It weighs about 1/50th of the total body weight and works constantly to maintain human health without any complaint. However, when overworked with excessive drinking, damage or cirrhosis of the liver can result, which is in fourth position among deadly geriatric diseases in Japan. The artificial heart is a substitute machine for the

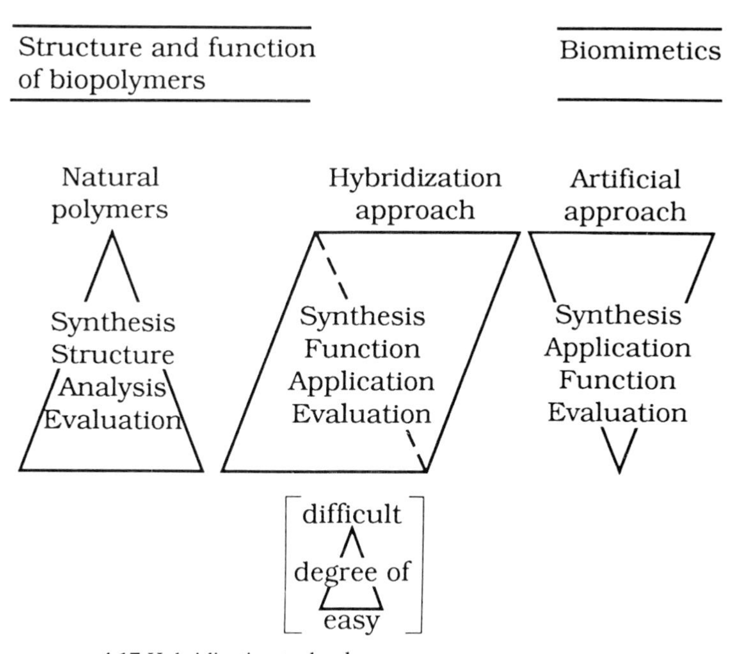

4.17 Hybridisation technology.

simple job of pumping blood, but the artificial liver is required to substitute for more than a thousand functions. Accordingly a hybrid-type artificial liver is the first step to producing a genuine artificial liver.

Liver cells lose their physiological functions and eventually die outside the living organ. Thus to develop a hybrid-type artificial liver, a base to fix and stabilise living liver cells is urgently needed. This base will be in direct contact and interact with living cells and the complex system of biomedical materials which envelopes the cells. These materials are specific for each particular cell. Professors T. Akaike (Tokyo Institute of Technology) and K. Kobayashi (Nagoya University) have developed a new technology to maintain living liver cells in a stable state outside the living organ. The success of this technology is due to the development of a new type of base material for fixing liver cells, which makes it possible to separate and/or recover a particular group of liver cells. Modified polystyrene was used for this base material, in which lactose was introduced as side chains attached to the polystyrene skeleton. This base material, unlike the model base of glycoproteins in the blood, adheres to living liver cells firmly for long periods, and also can detach from the cells when necessary (see Fig. 4.18). By developing this epoch-making base material further, it will be possible to build a device to analyse the response of cell functions (a biosimulator), or a system to function as a human organ (a hybrid-type artificial organ). Hitherto this has existed only as a dream in the field of biomedicine and cell technology.

The liver is known to generate body heat, to store blood, to synthesise vitamin A and to produce substances which reduce blood pressure or prevent

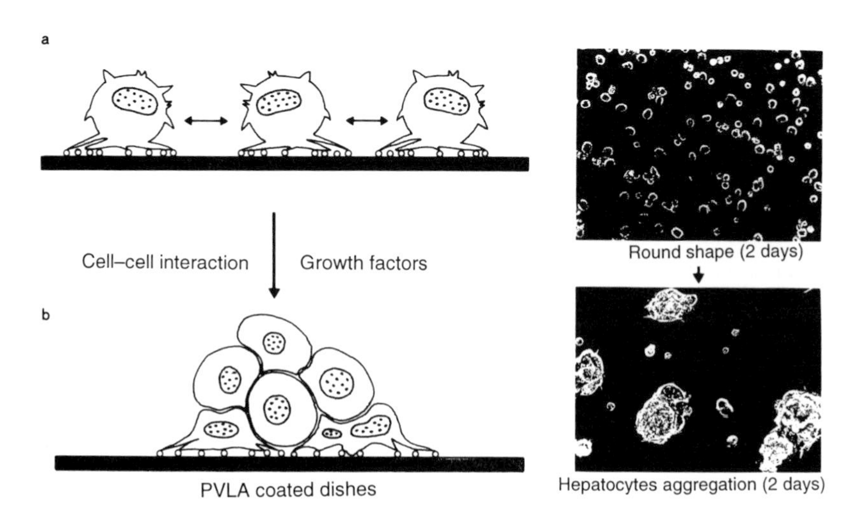

4.18 Hepatocyte aggregation on PVLA-coated dishes.

anaemia. These liver functions are conducted by liver cells. If we could understand the mechanism of drug metabolism by liver cells, the science of drug synthesis would be greatly advanced. Since liver cells can survive on this base material outside the living organ, this development also opens up a new system to investigate the metabolism of liver cells, and hence to develop new drugs. The artificial organ industry will attain a US$8 billion market over the next ten years. This polystyrene-based new material will now be applied to the cell-transfer and cell-separation base. The entire development bridges the boundary between materials and living processes.

4.2.2 A new treatment for incurable diseases: blood cell removal

Statistics indicate that mental stress and irritation, for example due to traffic jams, are the cause of an increased number of deaths. The experimental results of Professor K. Kataoka at the Science University of Tokyo show that the distribution of sub-groups of lymph cells changes when rats undergo stress, and consequently immunity is abnormally lost or improved. The immunity mechanism of human beings or other animals relates to two functions; natural healing power and the ability to combat an infectious disease. Vaccination to prevent whooping cough, diphtheria, Japanese encephalitis, influenza, tuberculosis (BCG vaccination), etc. utilises the second immunity mechanism where the acquired immunity works to protect from a particular disease. Although this immunological defence normally protects our health, it sometimes can cause serious indisposition, called self-immune response. In this type of disease, represented by chronic articular rheumatism and myasthenia gravis, the immunity mechanism recognises the constituents of the patient's own body as "foreign materials" and tries to eliminate them. No successful treatment is available at present. However, immune technology can be applied to these diseases, which would be cured if the antigen/antibody or their complex could be removed by filtration through an adsorbent material. The "lymphocyte removal treatment" was found to be effective with erythema or articular rheumatism by removing lymph cells that produce abnormal antibodies.

Professor Kataoka developed a new technological treatment, with the use of polymer materials, to improve the immune function by taking out lymph cells from circulating blood, removing abnormal cells chemically and returning them into the body system. The lymph cells play a central role in the immune mechanism, and are classified in many sub-groups of different functions. Among them are the B cells that produce antibodies, the T cells that regulate the immuno-response, the helper cells that strengthen the immunity,

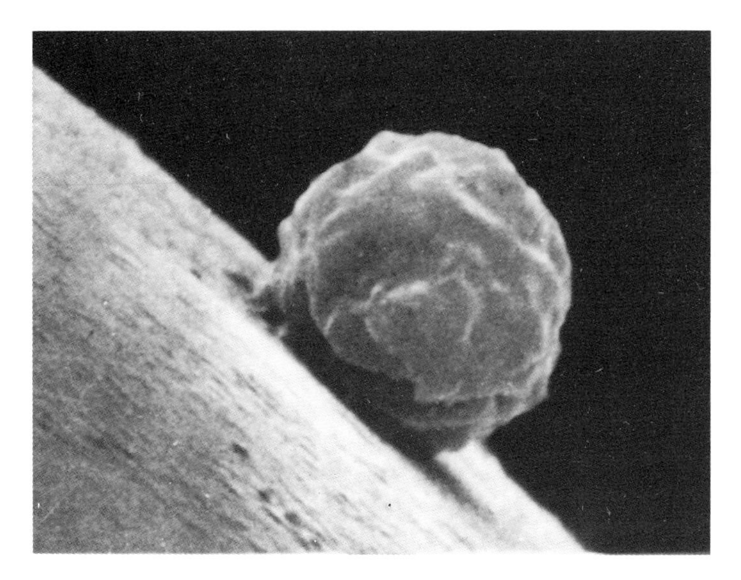

4.19 Lymphocytes (with their original spherical shape) adhered to the comb-shaped copolymer of hydroxyethyl methacrylate/polamine.

and the suppressor cells that suppress it. These cells are similar in terms of size and density, and cannot be separated by conventional ultracentrifugation. However, their membranes possess characteristic properties dependent upon the respective classes of cells on their surface. For example, immunoglobulin molecules penetrate from the surface of B cells but not from the T cells. Professor Kataoka utilised ionic interactions to distinguish between each class of cells. He developed an ion exchange resin (HA) that is able to separate two sub-groups of lymph cells (B and T cells) effectively in a short time. This ion exchange resin HA is made of hydroxyethylmethacrylate (HEMA) with polyamine side chains. The HA polymer has a micromosaic structure of polyamine islands in the HEMA sea. This structure is similar to that of the cell membrane, and discriminates between the B and T cells (see Fig. 4.19). B cells are selectively adsorbed in the pH range of 7.0–7.5 in a column filled with HA-coated glass beads, and T cells are separated in the effluents within 5 min in the range of physiological temperatures. Adsorbed B cells are recovered by simple stirring of the system later. This method can be applied for the separation not only of B and T cells, but also other cells that exhibit different isoelectric points, and is now used for the diagnosis of cancer or AIDS (acquired immune deficiency syndrone). In future, the analysis of lymph cells could be used in health control by identifying the change in the distribution of immune cells within our body system.

4.2.3 Iron-removal in sake by "fixed tannin"

Intrinsic functions, such as reactivity and affinity, can be introduced into functional materials, as exemplified by the "fixed tannin" developed by Tanabe Seiyaku Co. Fixed tannin is a fiber composed of cellulose with chemically coupled gallnut tannin, which selectively adsorbs and removes a particular component from a mixture of various biological materials. Fixed tannin technology was developed by the company, in collaboration with the National Research Institute of Brewing, and is now effectively used to remove iron ions from sake. The most important ingredients in brewed sake are water, rice and malted rice. If water with high iron content is used, brewed sake will be coloured and lose its flavour. Thus the iron content should be less than 0.02 ppm in water used for sake brewing. Persimmon tanning or active carbon was conventionally used to refine sake, but the long time taken adversely affected the subtle flavour of sake. Fixed tannin was found to be ideal in reducing the iron content of water for sake to less than 0.01 ppm and to refine sake, without any heating process and so maintain its freshness. This technology can now be applied widely in the fields of chemical industry, analytical chemistry, pharmacy and medicine.

4.2.4 New vaccine for hay fever

Pullulan is a water-soluble polysaccharide used as a food additive. Dr T. Matsuhashi, Senior Researcher at Okinaka Memorial Research Institute for Geriatric Diseases, is now developing a new vaccine for hay fever, using pullulan. Hay fever is caused by the immunoglobulin E (IgE) antibody in the body system. The new vaccine is made of pullulan coupled with antigen protein extracted from ceded pollen, which suppresses the formation of IgE in the body. This immunosuppressive drug is being developed in a joint project between the Institute of Public Health, Okinaka Memorial Research Institute and the Hayashibara Biochemical Laboratories Inc. The vaccine has been confirmed to be effective in the prevention of hay fever in animal experiments.

5 Biopolymer frontiers

Recently, interest has increased considerably in fiber-related biotechnology, and biofiber/biomedical materials, which can be grouped under the heading biopolymers. Here biopolymers denote the polymers that mimic or derive from natural organisms and exhibit similar functions to the natural material. These polymers now find application not only in the traditional areas of food and medical industries, but also in other industries, including information and communication. Biopolymers are thus an important ingredient in the rising generation of high-technology materials.

5.1 Mimicking the functions of enzymes and co-enzymes

Enzymes are essential in the fermentation of sake, soya bean paste, soya bean sauce and other foods used extensively in Japan. Life cannot be sustained without the enzymes which occur within the living structures of microorganisms, plants, animals and humans. Many of the chemical reactions of living systems, such as synthesis, decomposition, detoxification and energy supply, are controlled by enzymes. Enzymes are in most instances composed of spherical proteins, which exhibit their activity when bound to a low molecular weight compound, specified as the co-enzyme. The following examples show the particular applications of enzyme/co-enzyme electron transfer mechanisms.

5.1.1 Secondary fuel battery of synthezyme

Enzyme function is essential for the functioning of a stimulus-transferring nerve, a muscle contraction or a heart pulsation. The glow of a firefly is also enzyme-induced. Enzymes cooperate in a well-organised manner to perform a particular function, and often require metal ions such as iron and magnesium or other specific compounds in order to promote their action within the living body.

Cytochrome is a chromoprotein found predominantly in plant and animal cells, and is bound to iron porphorin to facilitate the transfer of electrons. Various attempts have been made to simulate the electron transfer mechanism of cytochrome proteins. A cytochrome film is as electrically conductive as germanium or silicon. Professor H. Shirai, in cooperation with Professor O. Hirabaru of Miyakonojo Technical College, synthesised a cytochrome-like enzyme and used this "synthezyme" to form the electrode for a secondary fuel battery. When an electric current is applied, water is decomposed into hydrogen and oxygen at the electrodes, although each component is seldom separated because of their rapid recombination. Shirai and Hirabaru used cobalt (instead of the iron in natural cytochrome) as the cooperative compound in their synthezyme, which was employed as an electrode on the oxygen side. The electrode operates by electron transfer, and oxygen at the electrode is catalytically reduced and stored by cobalt. This secondary fuel battery, with water as fuel and synthezyme as an electrode, is able to charge and discharge freely, and produce clean energy in the form of hydrogen, as schematically shown in Fig. 5.1. This synthezyme battery system could be developed further to replace lead batteries used in cars, or placed on a roof to harness solar energy to decompose water into

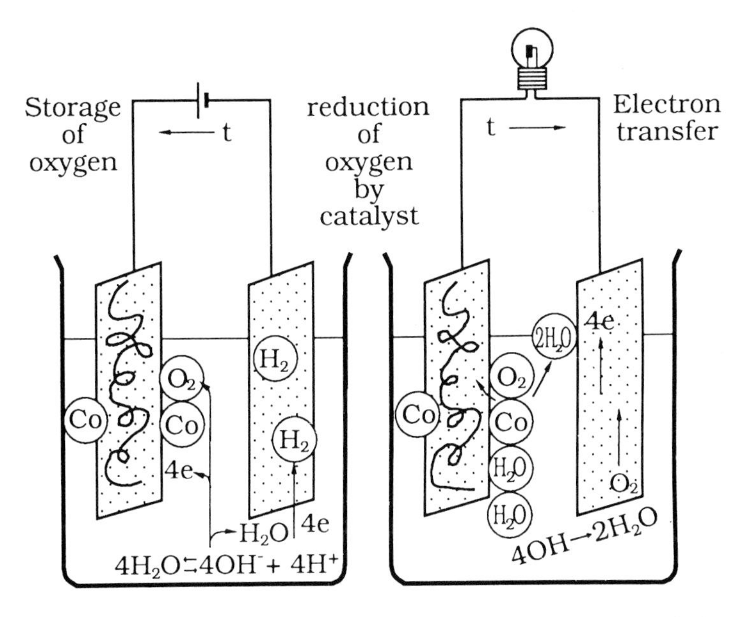

5.1 Secondary battery using synthezyme as electrode and water as fuel.

hydrogen gas, which could be used as a fuel to supply sufficient energy to heat a building, and consequently save natural resources. This synthezyme hydrogen-fuel battery could, therefore, provide a most efficient technology for storing energy by converting water into electric energy.

Biomimetics thus produced this secondary fuel battery, which can now be further improved in terms of energy efficiency and easy handling. The electron transfer mechanism of cytochrome can also be applied for the production of high-precision elements, such as on/off switches or memories operating at a molecular level. These high-precision elements are vital for the development of biocomputers, and are under continued investigation by the Ministry of International Trade and Industry, Japan.

5.1.2 Biobattery: An application of cell current

There are certain organisms that emit light or generate electricity. The glowing firefly is a familiar sight at the waterside on an early summer evening. *Cypridina hilgendorfii* glints mysteriously in the sea, and luminous *Armillariella mellea* or *Lampteromyces japonicus* surprises us in the mountains. All these light emissions are attributed to the action of enzymes. For example, a firefly glows when luciferin (a luminescent material) is oxidised by luciferase (a luminescent enzyme), where the heat produced by luciferin oxidation is converted into light energy by luciferase.

In 1791, Galvani observed a contraction of frog legs when joined by two metal pieces. Since then, much effort has been made to harness living organisms to produce electric energy. Advanced technology is necessary to study and harness energy or information stored within the living organisms. A hydrogen-fuelled biobattery is being developed using hydrogen-producing bacteria in the USA for use in the space shuttle.

Natural sources are being screened for suitable biosystems. For example, algae in the sea and lakes produce hydrogen by decomposing water with the aid of light energy. Blue-green algae grow in seawater by absorbing the sun's energy. Since the blue-green algae contain lipids, proteins and hydrocarbons, they can be used as an energy source as well as for the production of new bioactive materials.

An electric fish has a generator in its body. Its electromotive force varies from 30 V (an electric ray in the sea of Japan) or 400–450 V (an electric catfish in Africa) to 650–800 V (an electric eel in South America). The electric fish discharges intermittently as a means of self-defence or to catch food. It generates electricity through an ingenious use of the electroplax cells; its generation mechanism is similar to that of nerve excitation. A signal is transmitted from a synapse (a junction of the nerve fiber terminal to the next nerve cell) to an electroplax cell when the fish is excited. Then a particular side of the electroplax cell membrane depolarises and induces a potential difference on the other side of the membrane (see Fig. 5.2). Since many electroplax cells are stacked upon each other, a large potential is produced. This electric energy can be removed using an electrode, so providing a new way to access biochemical energy.

Electric fish are a special category in having electroplax cells, which are not found in most microorganisms and animals. However, muscle movement or nerve excitation produces a weak electric current, which can be monitored by an electroencephalogram or electrocardiogram. The objective of present research is to take out electric energy directly from the microorganism or animal cells. Although it is known that an electric current is generated when the cell is in contact with an electrode, the mechanism and materials involved in the current generation are not well understood. Recently, Professor T. Matsunaga, Tokyo University of Agriculture and Technology, identified the co-enzyme that transfers electrons from the cell. Electron transfer via this co-enzyme has been observed between an electrode and the cells of yeast, erythrocyte, macrophage and cancer cells. Since the maximum

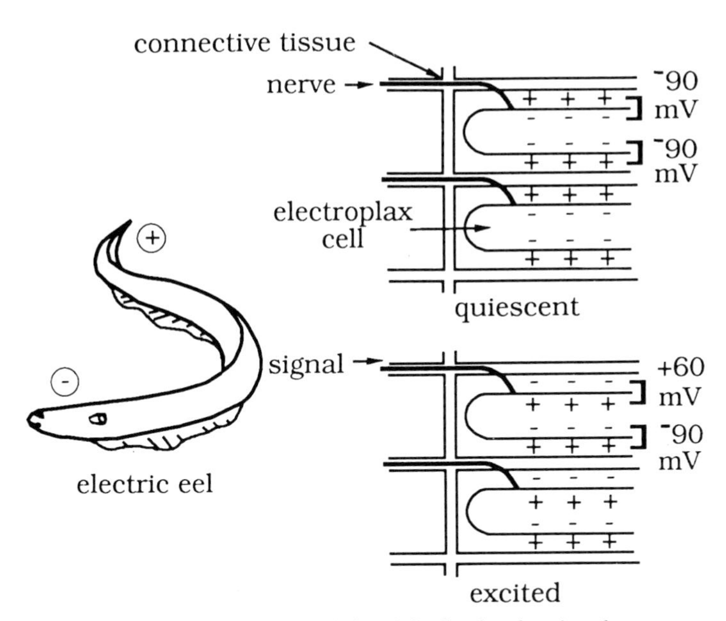

5.2 Mechanism of generation of electricity by the electric eel.

potential and potential peak shape of the generated current depends on the structure of the cell and cell wall, these can be used to distinguish the cell species, whether in microorganisms or animals. If an efficient method can be found to take out the electric current from the cell, it may be possible to develop a new type of microorganic battery, to interpret the cell information or to suppress the growth of cancer cells selectively.

5.2 Polysaccharides in semiconductors and medicine

Carbohydrates are composed of carbon, hydrogen and oxygen, and are found extensively in nature. Low molecular weight carbohydrates are normally referred to as sugars, and those of high molecular weight as polysaccharides, which can include simple polysaccharides and heteropolysaccharides, depending on the variety of constituent monosaccharide species.

There is a wide variety of monosaccharide species, classified according to the number of constituent carbon atoms they contain and the type of functional groups present. For example, glucose and fructose are monosaccharides found in honey and fruits, and are classified as hexoses, to denote that the monosaccharide contains six carbon atoms. Monosaccharides link together to form successively disaccharides, trisaccharides, tetrasaccharides and eventually polysaccharides. Cane sugar, or sucrose, is probably the most well-known disaccharide since it is consumed every day as a sweetener. Plant cellulose (used as cotton for making clothing) and starch (a food material) are examples of simple polysaccharides found abundantly in nature. Dry plants are made up of 85% polysaccharides and 15% other components. Cellulose makes up more than 90% of the composition of cotton, which is, therefore, referred to as cellulose fiber. Cellulose is produced not only by green plants, but also in large quantities by fungi and certain bacteria and in smaller quantities by sea algae and insects. Such ordinary polysaccharides are used for the physical protection of natural organisms, and in most cases have no specific functionality because they consist of a simple repeating structure.

Table 5.1 describes the simple polysaccharides, heteropolysaccharides and proteins found in nature. Although the heteropolysaccharides are present in small amounts, they can often be physiologically active. The study of the metabolism of polysaccharides was a forgotten subject for a considerable time, despite the fact that such studies could improve our understanding of life phenomena. Only recently have intensive investigations been undertaken in biochemistry, pharmacology, fiber chemistry, polymer chemistry and food chemistry to introduce a specific physiological function into simple polysaccharides.

Table 5.1. Polysaccharides and protein biopolymers

Field of science		Fiber science, polymer chemistry, food chemistry	Biology, biochemistry, pharmacology
Polysaccharides	Structure	Simple polysaccharides	Heteropolysaccharides
	Functions	Structural material such as cellulose, starch, pullulan and chitin	Mucopolysaccharides with pharmacological and physiological activities such as heparin and chondroitin sulphate
Proteins	Appearance	Fibrous	Spherical
	Functions	Structural proteins such as silk fibroin, collagen, and wood keratin	Functional proteins such as enzymes
	Properties	Large amounts, lack of physiological functions	Small amounts, high functionality
Reference		Natural fiber	

5.2.1 Improvement of integrated circuits using water-soluble natural polymers

ICs (integrated circuits) can be improved by printing finer circuits on to a silicon base. Matsushita Electric Industry Co. and Hayashibara Biochemical Laboratory developed a new type of photosensitizer to process the fine pattern for circuits of very large-scale integration (VLSI) of the next generation using the water-soluble natural polysaccharide pullulan. Pullulan has a similar structure to the edible polysaccharide starch, and can be removed easily by washing with water; it is not soluble in organic solvents.

The VLSI circuits are produced by photoprinting, that is, by exposing the photosensitive resin coated on the silicon base to ultraviolet radiation. The beam should be as fine as possible to improve the IC capacity. IC integration can be reduced to the submicrometre region when the photosensitive resin is coated with a thin layer of water-soluble photosensitiser. This functions as a filter and prevents beam scattering (see Fig. 5.3). Existing techniques can thus be used to produce 4 Mbit and even 16 Mbit DRAM (dynamic random access memory) by applying such new water-soluble photosensitisers. Additionally, production yields also improve, so that there is no need to introduce new sophisticated systems such as X-ray beams or electron beams. This application illustrates the value of cooperation between two industries in quite different fields, which resulted in a completely new application of natural polymers.

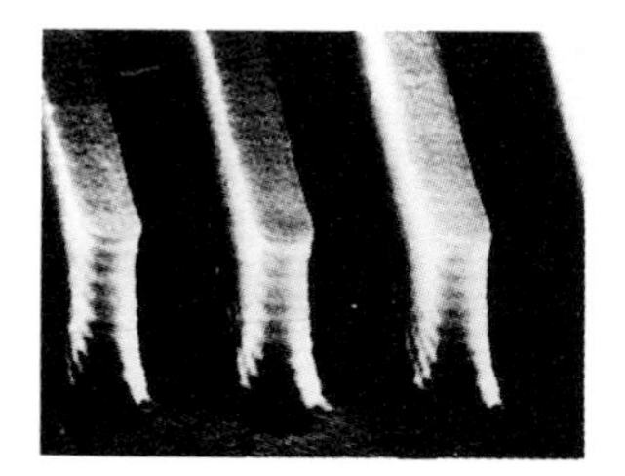
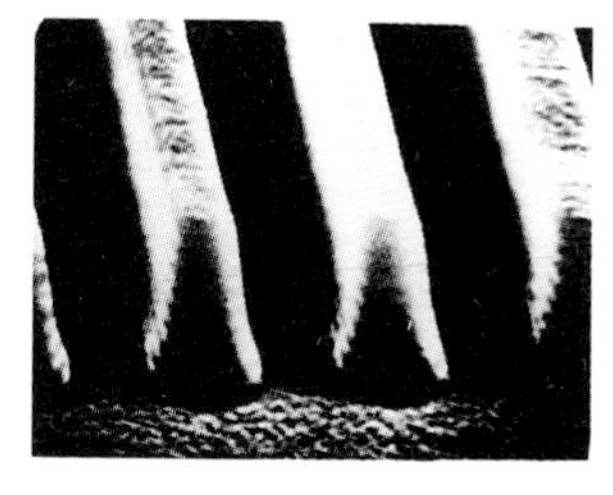

5.3 Scanning electron micrographs showing patterns with line width of 0.6 μm. Conventional photosensitive resin (right) and new water-soluble sensitiser (left).

5.2.2 Antitumour functions of polysaccharides

Plants and animals are provided with built-in mechanisms to survive, which are still beyond our complete understanding. As we discover more about such mechanisms, these find applications in high technology.

Polysaccharides, proteins and nucleic acids are three major biopolymers, which sustain the physiological functions of living organisms. Among them, polysaccharides are known to play an important role in the division of cells and their subsequent growth, while preserving cell characteristics. Mushrooms produce polysaccharides of complicated structure. The Japanese have been aware that mushrooms can be used in an effective anticancer therapy for some time. Two antitumour drugs appeared on the market from this source, namely lentinan and schizophyllan. These are produced from extracts from *Lentinus edodes* (Chinese mushroom) and *Schizophyllum commune*, respectively. Lentinan is a polysaccharide of white powdery appearance with molecular weight about 500,000. Research workers at Ajinomoto Co. Ltd found that lentinan is made up only of β-1, 3-linked glucans. They were found to be physiologically active during the course of the study of structure in relation to anticancer activity using various polysaccharide derivatives. Lentinan has a complicated three-dimensional structure, and exhibits its anticancer activity towards a particular type of cancer as well as exhibiting synergic effects with other chemical anticancer agents. Moreover, it does not show the adverse side-effects that are characteristic of many antitumour chemicals.

The triple-stranded structure of schizophyllan (see Fig. 5.4) is responsible for its antitumour activity. It is a polysaccharide composed solely of glucan bases, and Professor N. Norisue of Osaka University found that antitumour activity is induced when the molecular weight exceeds 50,000 and the triple-stranded structure is formed. Schizophyllan is now commercially available from Taito Co. Ltd. It is extracted from cultivated *Schizophyllum commune*, depolymerised and purified before use by inflection.

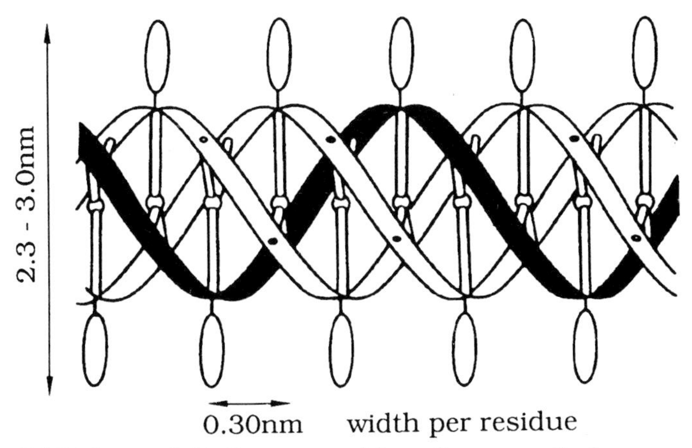

5.4 Triple-stranded helix structure of the antitumour medication Schizophyllan.

Taito Co. and Ajinomoto Co. have employed different types of mushrooms to develop antitumour agents, but came to the same conclusion that polysaccharides with β-1,3-linked glucan structures possess the antitumour activity. The anticancer agents derived from mushrooms may not attack cancer cells directly, but increase the immunopotential of the body to suppress cancer cell growth. Activation of the immune response is due to cell-mediated immunity through macrophase, cytotoxic T cells and natural killer cells, rather than by tumoral immunity. The relationship between polysaccharide structure in solution and physiological activity now requires further intensive investigation.

The Mizuno Biohollonics Project (led by Professor D. Mizuno of Teikyo University), within the framework of the Creative Science and Technology programme organised by the Research Development Corporation of Japan, has already identified the potential of activated macrophages in cancer treatment. As yet there have been no clinical use or commercial manufacturing process developed. The Japanese Government has just started a ten-year project in this area to develop new cancer treatments. Considering their physiological role within the cell, polysaccharides need to be investigated more intensively in various fields of medicine, pharmacology, biology, chemistry, physics, and even fiber science and technology. In this context there is a need in Japan for a National Saccharide Research Institute, where researchers with different backgrounds can cooperate.

5.2.3 Polysaccharides as antitumour-agent carriers

The cells making up an organism originate from a single fertilised egg cell, which on repeated division yields cells specific to a particular organ, such as liver, heart, skin or nerve, so differentiating their functions. The polysaccharides on the cell surface are believed to be involved in the mechanism of intercellular interaction.

Although dextran, a polysaccharide produced by a lactic acid bacteria *Leuconostoc mesenteroides*, is pharmacologically inactive, it can be used as a carrier for antitumour drugs, according to the results of Professor M. Hashida of Kyoto University. It is an α-1,6-linked polyglucose. An antitumour drug should remain in the blood at a certain concentration range for a considerable period if it is to work effectively. This is often a problem since the drug concentration usually drops sharply within a short period. If an excess of drug is administered, undesired side-effects can appear. Dextran when used as a carrier of the antitumour drug maintains the drug concentration at a desired level over long periods. This facilitates drug action and suppresses its side-effect.

Professor J. Sunamoto is now investigating the possibility of using *konjak-mannan*-coated liposome as a drug carrier. *Konjak mannan* is a hetero-polysaccharide composed of β-1,4-linked mannose and glucose, and is used regularly in foods in Japan. Since phagocytes tend to encapsulate the *konjak-mannan*-coated liposome which concentrates specifically in the lungs, this system can be applied in the treatment of metastic lung cancer. The

5.5 The sea hare (*Encylopedia on Animals:* Japan Mail Order Co.).

polysaccharide-coated liposome can also be used as a drug delivery system to target particular organs or tissues.

5.2.4 Protective mechanisms of the sea hare (Aplysia kurodai)

A sea hare (see Fig. 5.5) grows without a hard shell, and secretes purple sweat to protect its soft body when an enemy approaches. Its egg survives without being infected by bacteria or being eaten by other animals. Professor J. Mizuno of Teikyo University is developing a unique antitumour drug by using a glycoprotein found in the eggs of the sea hare. This protein is believed to participate in its bioprotective mechanism.

The glycoprotein containing 10% saccharide was isolated from sea hare eggs, and was found to destroy cancer cells selectively in mice and humans. This example illustrates that living systems utilise many mysterious materials to achieve survival. Saccharides involved in cell recognition and immunity can surely assist in our own control of physiological functions.

5.3 Biomass of crab and shrimp shells

Chitin, a polysaccharide, is the major component of crab and shrimp shells. Its structure is similar to that of cellulose. Chitin was not considered of value until physiologically active materials were derived from this source. Professor S. Tokura of the Hokkaido University initiated the chitin-utilisation project in 1972, at the request of the Hokkaido Prefectural Office. Professor S. Hirano of Tottori University also started his investigation of the physiological activities of chitin and chitosan in the early 1970s. These two have contributed extensively to this subject since that time.

Chitin has been shown to be physiologically active in relation to an antithrombus effect and adjuvant activity. It can also be used as a food material to control cholesterol levels in blood. The application of its derivative chitosan is now being widely developed: as a slow drug-release membrane; colour adsorbent; antifriction agent for paper; in cosmetics such as the softener/moisture-retainer (conditioner) for hair; drug carrier in the form of porous beads, and as an enzyme immobiliser.

5.3.1 Artificial skin from crab and shrimp shell

Crab meat tastes especially fine in winter, but its shell was simply a disposal problem until recently. Crab and shrimp shells are now considered second only to cellulose as a useful biomass. Since chitin is decomposed and

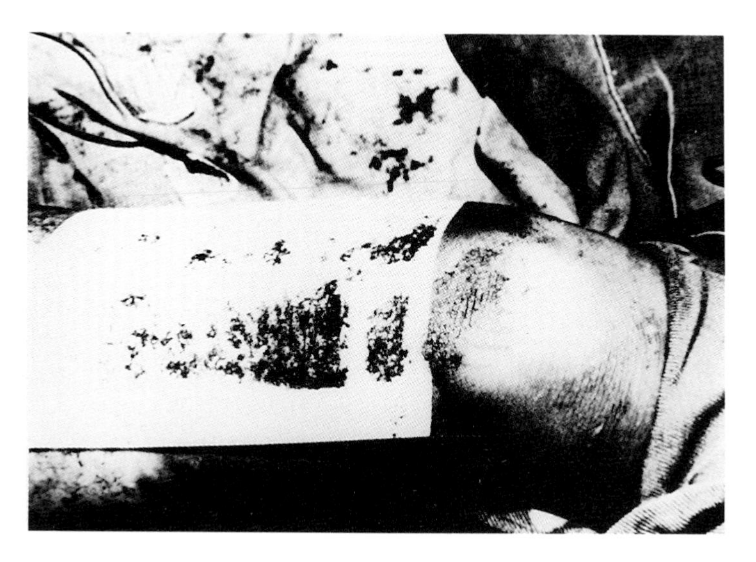

5.6 Artificial skin derived from chitin (Unitika Ltd).

absorbed in the living organism, it can be used for medical applications. Unitika Ltd developed an artificial dressing, Beschitin-W (see Fig. 5.6), from crab shells. Crabs are extensively fished around Florida and California, USA. Dr Austin, of Delaware University near Florida, holds the basic patent on chitin fiber production. Unitika established an industrial process to produce chitin fibers of various diameters, based on this patent. The choice of solvent is critical and the dissolving conditions must be carefully controlled in the process.

Fibers, fabrics, sutures and non-woven fabrics from chitin are now available. For example, chitin non-woven fabric has excellent characteristics as an artificial skin because of its good adhesion to the human body surface and its value in stimulating new skin formation. It promotes no immune antigen–antibody reaction, and substantially accelerates healing and reduces pain. Unitika started the commercial sale of chitin products in April 1988, in collaboration with Roussel Medica Co. (a subsidiary of the medical marketing company in France). A standard type artificial skin (10 × 12 cm) costs more than $200 at present.

5.3.2 Porous chitosan beads

Alkali treatment of chitin yields chitosan, which is now allowed as a food additive. Since chitosan dissolves easily in acetic acid (vinegar), its processing has been widely investigated for the production of chitosan fiber or film, which can be conveniently applied in the food and biomedical industries.

Professor S. Tokura developed a continuous spinning process for chitosan in 1982. It can be spun in a long filament of a similar strength to rayon, although the present process yields a somewhat thicker filament than cotton. Fuji Spinning Co. Ltd have developed chitosan porous beads, consisting of numerous pores running radially from the surface to the centre. The active surface area, including the interior pores exceeds 200 m^2/g. In other words, the total surface area of 50 g chitosan beads is approximately equal to the ground area of the Korakuen Dome or Wembley Stadium. Because of their good compatibility with living cells, the beads are used for cell culture, and do, in fact, promote cell proliferation. Chitosan porous beads can also be processed into fine beads 10 μm in diameter and the Fuji Spinning Co. now supplies these find beads for HPLC (high-pressure liquid chromatography) in association with Showa Denko Co. Chitosan beads can also be used as a drug carrier, or as a slow drug releaser.

5.4 New applications of silk

Silk has been used as a thread after being spun by silkworms, and in Japan nobody dared, for centuries, to trespass upon the territory of holy silkworms by reprocessing silk threads. As time passed, however, silk lost this aura, and is now used in many forms such as thread, film and powder in order to add to it more commercial value. The raw silk thread produced by a silkworm is composed of fibroid in the core and is covered with sericine. If the sericine is removed by degumming, water-insoluble fibroin remains as a fine silk thread. This thread has a triangular cross-section and, therefore, exhibits a characteristic lustre, pleasant handling and an elegant drape. Silk was also found to be biologically active and Professor T. Asakura specified the structural and reactive regions within silk fibroin molecules. The following examples will demonstrate recent applications of silk in the field of biotechnology.

5.4.1 Stockings of hybrid silk

"Silk" induced a special nostalgia, nobleness and elegance in past generations. It has been the Queen of Fibers ever since the days of Silk Road. The mysterious processes that take place after the silkworms eat mulberry leaves to produce silk fibroid have gradually been unravelled. Recently its good biocompatibility properties opened up new areas of application as a biomaterial for sutures and for enzyme-immobilisation.

Nylon is the synthetic fiber that was introduced to copy silk, and to replace it for use in ladies stockings in the 1950s and 1960s. Now silkworms can produce finer and longer silk than nylon using biotechnology. The National Institute of Sericultural and Entomological Science, Japan, has succeeded in breeding a new type of silkworm that can produce a fine homogeneous silk filament of about 1,500 m in length. The National Institute has developed the use of this silk in cooperation with the Asahi Chemical Co., Professor A. Shimazaki (Shinshu University) and Professor N. Naruse (Bunka Women's University). This hybrid silk was commercialised under the trade name of Silran in the autumn of 1987. It is produced by extruding silk and nylon together through an air-jet nozzle with nylon filament placed at the centre and five raw silk filaments (2 denier each) twined around the nylon core. Thus the hybrid silk is composed of fine silk at the surface of a synthetic fiber core. It thus retains excellent handling and a good silk lustre as well as having the fiber strength of nylon. When the stockings from this hybrid silk are degummed to remove surface sericine, they show a metallic-silver lustre. The fibroid threads also bulk-up because of the thermal shrinkage of the nylon core. This product is, therefore, a genuine hybrid of silk and nylon, and yields quite different characteristics from the conventional blended yarn of silk and nylon. The wearing test of such hybrid silk stockings demonstrated its good handling, and such products are now commercially available.

5.4.2 Shape-memory silk yarn

A highly elastic silk yarn was developed by S. Mizushima of Mizushima Silk Industry Ltd and commercialised by Daito Spinning and Weaving Co. Ltd. Although both yarns are made of protein, a silk yarn lacks elasticity in comparison with wool because of the difference in internal structure. In the process, silk yarn is chemically treated by dipping in a solution of hydrolysed fibroid keratin and collagen, dried, crimpled, dipped again in water and thermo-set in the wet state under high pressure (2–3 atmospheres; 200–300 kPa) at 110 °C for 10 min, yielding shape-memory silk (see Fig. 5.7). When this product is wet-heated at 60 °C, the silk yarn becomes crimpled and bulky. Since its twisted structure is fixed in the memory even when the silk is untwisted again into uncurled yarn, the silk yarn reversibly recovers its curled shape by steaming. This elastic silk yarn can be applied in various textile products including outer garments, tights and knitted yarns.

5.4.3 Biocosmetics

Biotechnology has now provided a way to produce new products from silk having specific functions. Kanebo developed the facial treatment cosmetics Fresh-up Powder and Bio-Powder Foam from silk in order to capture the

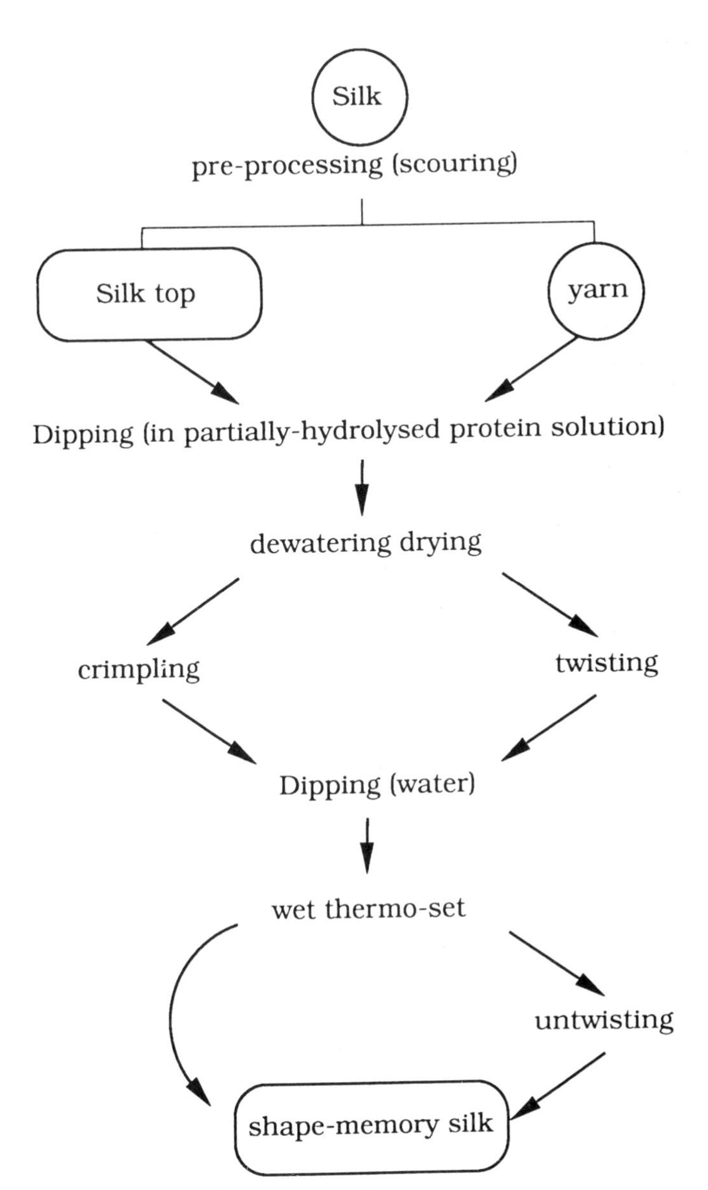

5.7 Processing of shape-memory silk.

elegant image of silk in the market. Fatty dirts on the skin surface can be removed by washing with soap. Protein dirts from dead skin, however, cannot readily be removed unless they are first hydrolysed with the enzyme proteinase. Certain of the conventional cosmetics contain proteinase, but its hydrolysis activity deteriorates rapidly with time, particularly when left with surfactant in the wet state. Many attempts have been made to maintain the proteinase activity in cosmetics. Kanebo applied an enzyme immobilisation technique for this purpose by encapsulating proteinase in water-insoluble fibroin. The fibroin regenerated from aqueous solution has similar characteristics to liquid silk in the silkworm body. Fibroin will crystallise easily and become water-insoluble when salt or alcohol is added, dried or stirred. Kanebo developed the technology to produce fibroin film, powder, fiber or gel by varying the conditions of crystallisation from aqueous solutions. When processed with added enzyme or antibody in the solution, the fibroid crystallises with the enzyme or antibody trapped inside. This process can be applied to immobilise the enzyme, despite the fact that direct immobilisation of proteinase with fibroin was not possible. The pH value is controlled to suppress the enzymic activity of proteinase, and the proteinase solution is added to the fibroin solution. Fibroin powder containing proteinase is salted out from the mixed solution. The encapsulated proteinase is stable to heat and its hydrolysis activity lasts for a considerable period, since fibroid protects proteinase against heat and moisture. The enzymic activity deteriorates only 10% after 300-day storage at 45 °C in the fibroin. Fresh-up

5.8 Scanning electron micrograph of proteinase-encapsulated fibroin powder.

Powder is made of granules of mixed proteinase-encapsulated fibroin powder and detergent (Fig. 5.8). The same know-how can be applied to prepare biosensors, drug carriers and bio-reactors for food.

5.5 Fibers produced by bacteria

5.5.1 Bacterial cellulose

Cellulose, as noted previously, is the main component of cell walls of plants. The cellulose content of wood pulp is 50 to 60%, and as high as 90% in cotton. It has been known for some time that some bacteria, mostly acetic acid producing bacteria, also produce fibrous cellulose without the aid of light. This bacterial fiber is very pure cellulose, and many investigations have been undertaken to identify the bacterial strain and cultivate it effectively. For this Ajinomoto Co. selected *Acetobactor aceti*, the acetic acid bacterium, 1 × 2 to 1 × 3 mm in size. When cultivated for seven to ten days in a medium containing 5% sucrose, nitrogen and salts at 30 °C, this bacterial strain

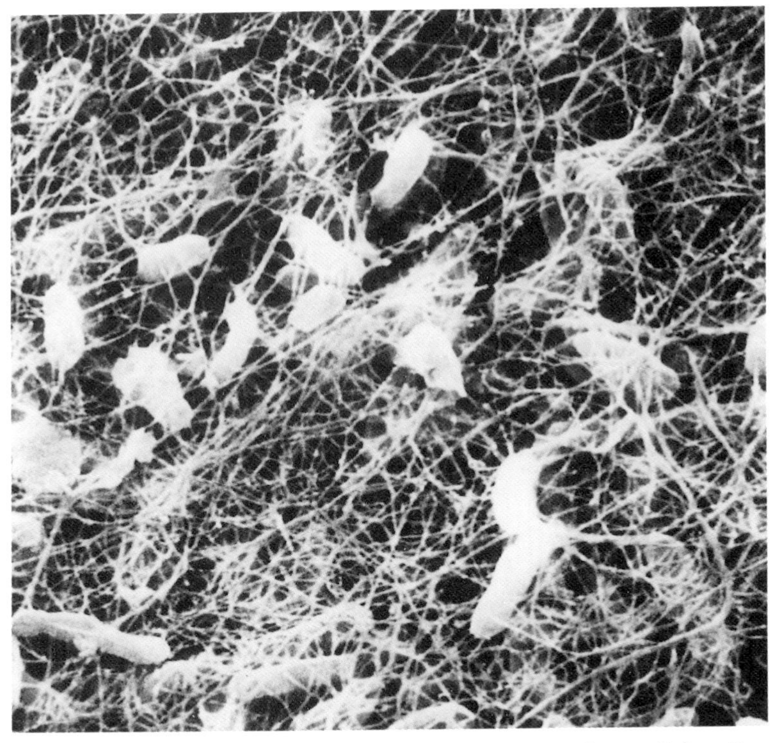

5.9 *Acetobacter aceti*, an acetic acid bacterium, produces cellulose fibers outside the organism (Ajinomoto Co. Inc.).

produces a gel-like material containing fine cellulose fiber, which is too thin (about 20–50 mm in diameter) to classify in terms of the conventional denier unit (see Fig. 5.9). This fiber has great potential for paper, medical and other industrial applications as a new functional material. In 1997 it was introduced in food products by the Kelco–Nutrasweet Company.

It is not yet fully understood how the bacteria produces the fibrous cellulose; the mechanism is of great interest from a scientific viewpoint. Fine gel-like fibers are spun from holes of several nm in diameter on the bacterial cell surface, to form ribbon-like gel sheets outside the cell (see Fig. 5.10). These sheets can be processed into paper of extremely high Young's modulus (30 GPa), which is almost equivalent to that of aluminium. The sheets cannot be torn easily by hand. These unique characteristics of the bacterial cellulose sheet are due to (i) a high degree of cellulose crystallinity; (ii) a high lateral order of the crystallites, (iii) a lamellar alignment of crystallites in several layers; and (iv) a network structure of the sheet.

Ajinomoto, the Research Institute for Polymers and Textiles and Sony are jointly working to develop new applications for the bacterial cellulose sheets. Already Sony has employed these sheets for loudspeaker diaphragms, in order to capitalise on their high specific modulus and high internal loss. The sound reproducibility is outstanding, and Sony has now commercialised high-quality headphones using these bacterial cellulose sheets (see Fig. 5.11).

The extremely fine filament of bacterial cellulose could realise many scientific dreams for textiles. For example, its application is now being extended to produce a new type of artificial leather with a mild touch.

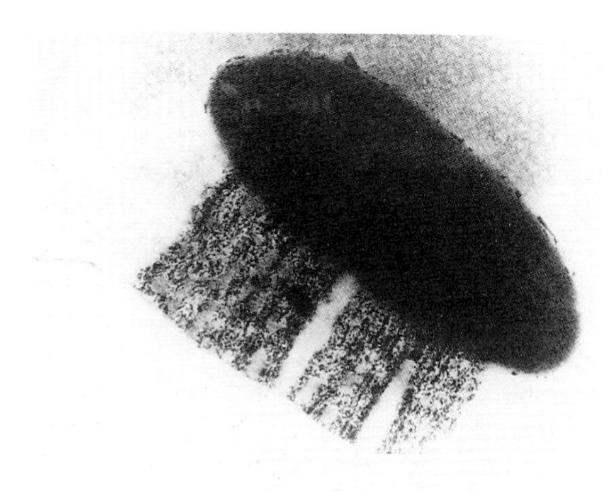

5.10 *Acetobacter aceti* produced ultra-fine cellulose: magnified version of Fig. 5.9 (Dr A. Kai, Tokyo Metropolitan University).

5.11 High-quality headphones of bacteria cellulose (Sony and Ajinomoto).

5.5.2 Bacterial polyester as strong as nylon

As *Acetobacter aceti* produced cellulose, so have bacteria produced polyester for several hundred million years. More than a hundred bacterial species are known to be polyester-producing, which include *Alcaligenes sp.*, *Bacillus sp.*, photosynthetic bacteria and blue-green algae. These microorganisms produce and store polyester, which can be used as an energy source in case of starvation, in the same way as animals and plants store energy in the form of glycogen and amylopectin, respectively (see Fig. 5.12).

The polyester so produced is stored in the bacterial body as particles of 0.5–1.0 μm in diameter, which can be extracted using organic solvents. Natural polymers such as cellulose and fibroid do not melt, but natural polyester is exceptionally thermoplastic and melts at about 180 °C. It can, therefore, be moulded into any shape like any other synthetic polyester.

The polyester-producing bacteria contain both polymerase (an enzyme controlling polymerisation) and depolymerase (an enzyme inducing depolymerisation). Thus, bacterial polyester is biodegradable. Bacterial polyester is, therefore, expected to be used where a natural hazard is created by non-degradable rubbish. It could be an important material in establishing a well-balanced ecosystem. Since the bacterial polyester is composed of highly crystalline D-3-hydroxybutyrate, the product at present is too stiff and brittle for major practical use. Recently a new method was developed to produce bacterial copolymeric polyester efficiently by fermenting a suitable combination of bacteria and food. ICI, UK, has applied this fermentation

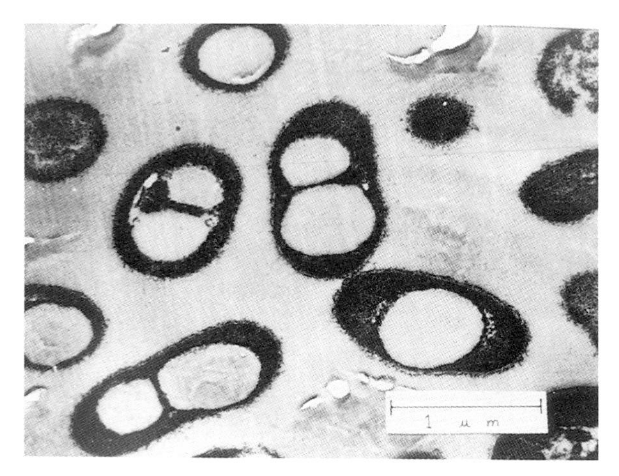

5.12 Polyester stored in *Alcaligenes eutrophus* (scanning electron micrograph by Prof. Y. Doi).

technology to produce a random copolyester of 3-hydroxybutyrate (3HB) and 3-hydroxyvalerate (3HV) by feeding these two ester carbon sources to the bacteria. This copolyester can have its 3HV component varying in concentration from 0 to 30% and is now commercially available under the trade name of Bipol. The price will decrease to less than US$6/kg if the annual polymer production exceeds several thousand tonnes.

In 1987 Professor Y. Doi, Research Laboratory of Resources Utilisation, Tokyo Institute of Technology, found another combination of bacteria and food that can produce a copolyester by fermentation. *Alcaligenes eutrophus* produces a copolyester composed of 3-hydroxybutyrate (3HB) and 4-hydroxybutyrate (4HB) when fed with 4-hydroxybutyric acid. The physical characteristics of these bacterial copolyesters vary from elastic rubber-like to hard plastic-like appearance as the proportions of 3HB and 4HB are varied. For example, this bacterial copolyester can be processed into a fiber that is as strong as nylon in terms of tensile strength.

Bioplastics, like the bacterial polyesters, are biodegradable, and these materials can be decomposed by bacteria in sludge and soil. The biodegradability and hydrolysis characteristics can be controlled by changing the copolymer composition and molecular weight. A slow releasing system for agricultural chemicals is now being developed using biodegradable polyester microcapsules containing chemicals. These microcapsules are decomposed in the soil gradually, with the speed controlled by copolymer composition and/or molecular weight, and thus release chemicals over a long period.

Since the biopolyesters are biocompatible they also find applications in medicine. The surgical suture, gauze, bandage or the material used to repair bone fractures or deficiencies made from bacterial polyester cause no inflammation in the organs or tissues where they are applied. Since these biopolyesters are optically active and piezoelectric, further applications may be anticipated in the field of optics and electronics.

5.6 New functions for cellulose

The solubility and hydrophilic/hydrophobic character of cellulose can be readily modified according to the application required. Consequently, cellulose can be used industrially as an additive to drugs and cosmetics, or as a coating material for tablets. Specific functionality can also be introduced by chemical modification of cellulose (see Fig. 5.13). In this section, selected examples will demonstrate the novel functions of cellulose developed recently.

5.6.1 Absorption of mutagen by Blue-cotton

The change in eating habits and environmental factors have caused an increase in bowel, pancreas and lung cancers in Japan. Rapid Europeanisation, particularly the introduction of new foods after World War II, has resulted in a change in the physical characteristics of individual Japanese. Moreover, air pollution due to rapid industrialisation has led to an increase in particular types of cancers. A simple cancer diagnostic test has been developed, which allows the factors causing cancer to be analysed in detail. It requires enormous time and money to prove scientifically that certain materials are cancer-causing (carcinogenic). The rapid use of a mutagenic test is, therefore, extremely useful as a first screening programme. Such a test does not positively prove that the material is a carcinogen. However, it allows possible offending substances to be quickly identified, and then subjected to more intensive investigation.

The development of this test is interesting. Professor H. Hayatsu, Faculty of Pharmaceutical Sciences, Okayama University, found a mutagen in the pigment of blue chalk used every day in his university. Although the pigment itself is not a mutagen, it eventually transpired that copper phthalocyanine (blue chalk pigment) is easily contaminated with a mutagen during the chalk production process. It was evident that the copper phthalocyaninic pigment absorbs mutagens selectively. This finding led to the invention of a new material "Blue-cotton", which is currently used for the simple mutagenic test.

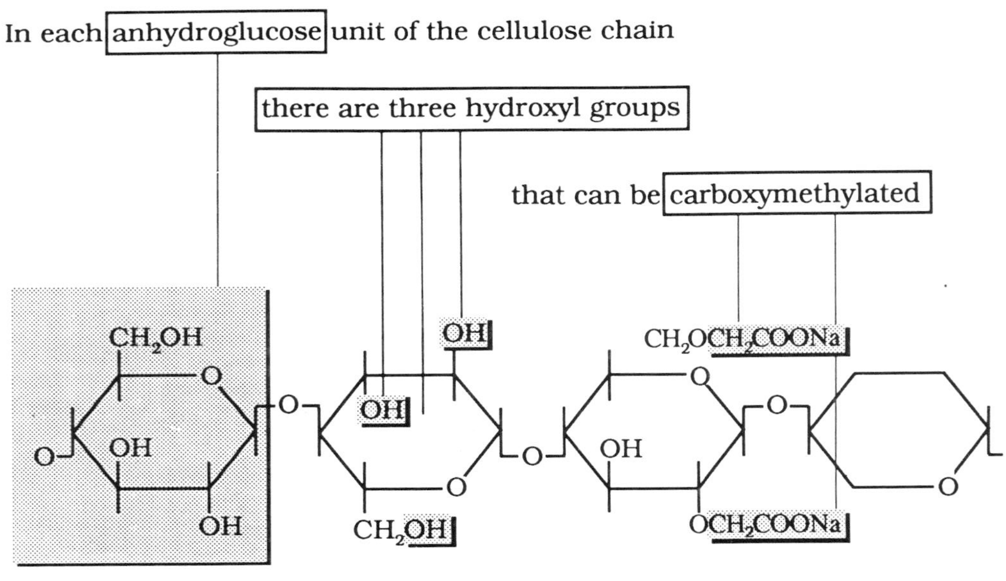

5.13 Possible sites for chemical modification of cellulose.

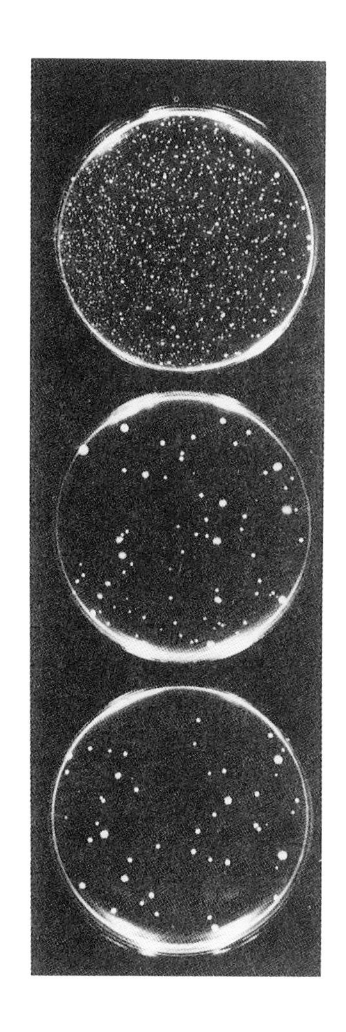

5.14 Absorption mutagen by Blue-cotton: Trp-P-1 mutagen (carcinogen).

"Blue-cotton" is named after its appearance, and consists of cotton fiber stained with copper phthalocyanine, a widely used blue pigment. Grilled fish or meat contains strong mutagens, but these mutagens do not readily dissolve in water. Professor Hayatsu wished to identify mutagens dissolved in water, and for this reason developed "Blue-cotton". The mutagens in water (normally in ppm) are adsorbed on to cotton fiber and form a complex with the copper phthalocyanine (Fig. 5.14). Those mutagens can be recovered using alcohol/ammonium aqueous solution, and can, thereafter, be submitted for further identification tests.

"Blue-cotton" is produced and distributed by Sumitomo Chemical Co. and is used now not only for the mutagenic test, but also as the basis for exotic clothing materials.

5.6.2 Exclusion of mutagens from the body

Asparagus cellulosics adsorb certain carcinogenic substances and can prevent them from being ingested by the body. This was discovered by Mr. T. Emura, Technical Adviser of Hokkaido Local Council, and the National Institute of Gene Research, Mishima, in a research project commissioned by Kamiwakubetsu Town, Hokkaido. Kamiwakubetsu Town produces asparagus, and ships it to the markets after eliminating the hard stems. Such stem wastes amount to 300 tonnes every year, and research was started in 1984 in an attempt to find some use for the asparagus wastes.

The cellulosic component is conventionally separated by treating the asparagus stems with acid or alkali, when the other components including hemicellulose are decomposed. A new process was developed that consists of cooking the asparagus stems, rinsing and drying. In this way most of the asparagus components (some proteins, hemicellulose, pectin, amylopectin, calcium and iron) remain undecomposed and are trapped within the cellulosic network. The asparagus cellulosic network was shown to adsorb mutagens, such as tar components in cigars and benzopyrenes, and keeps them from being incorporated in the human body.

Kamiwakebetsu Town constructed a processing factory for asparagus stems in 1987, and now sells the asparagus cellulosics as a health-promoting food through Yoshitomi Pharmaceutical Co. This cellulosics product was first publicly introduced at the Exhibition of Hokkaido Technical Research Institute held in February 1987, and attracted much attention. Only 6 tonnes of asparagus cellulosics can be produced by processing 300 tonnes of asparagus stems. Asparagus stems are now collected in the entire Hokkaido area in order to extend future asparagus cellulosics production.

5.6.3 Fiber as food

Dietary fibers are now accepted to be an important ingredient of a healthy diet, and are thought to reduce the probability of cancer induction in the digestive organs. Dietary fibers shorten the time food remains in the digestive organs, and as a result a carcinogen will remain in contact with these organs for a shorter period. Dietary fibers are believed to suppress the secretion of certain components in bile fluid that might induce colon cancer. Moreover, it is necessary to chew food better in such a diet, so stimulating the secretion of digestive juices. It has also been shown that food with a higher content of dietary fibers, while staying for a shorter time in the body overall, remains a relatively longer time in the stomach, so creating the sensation of satisfaction longer. A daily consumption of 20–35 g of dietary fiber is said to be necessary for good health, according to the individual's weight. Cellulose is the main component of dietary fiber, and is not significantly metabolised by the body, and as a result is not weight-inducing either.

Microcrystalline cellulose powder is accepted as a food additive. However, when the cellulose content exceeds a few percent of this processed powder food, the taste deteriorates almost to the point of being unedible. Asahi Chemical Industry Co. consequently developed a new food additive Sekicel from cellulose, which is pleasant to eat. Sekicel is a fibrous material spun from a homogeneous solution of alkali-soluble cellulose and starch. Cellulose generally does not dissolve in aqueous alkali solution, but swells as a result of its inter- and intra-molecular hydrogen bonding. When high-pressure steam is applied to cellulose, such hydrogen bonds are broken, resulting in powdery cellulose becoming soluble in alkali. Asahi Chemical Industry Co. also found that aqueous sodium hydroxide solution dissolves cellulose, specifically when the alkali concentration is about 9% by weight. Other alkali-soluble materials (such as protein or polysaccharide) can be mixed with cellulose in this alkali medium which can then be processed into a shaped material product. Sekicel so spun from the mixed solution of cellulose and starch has a porous texture, with pore diameter about 1 μm (see Fig. 5.15). Two types of Sekicel product are available commercially. One is a wet type of cut fibers and the other a dry type of milled fibers. Sekicel is a white food additive with no taste nor smell, has a pleasant and soft touch to the palate in spite of its high cellulose content. It retains water and oil well, and exhibits good stability against boiling. The food can first be processed into a solid shape and becomes dietary by adding Sekicel. Sekicel is mostly used in processed foods such as pastry, with flaky, powdery or granular appearance, and in replica foods such as artificial scallop, crab meat and shark's fin (which contain 10–50% Sekicel) and ground pollack meat. Sekicel now attracts much attention as a functional food additive.

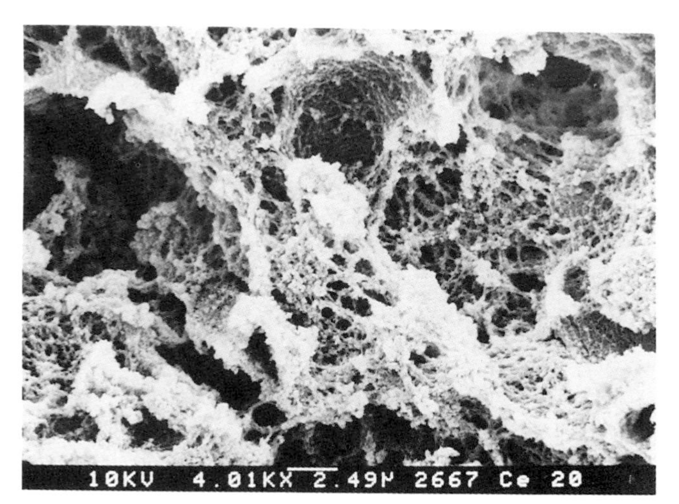

5.15 Sekicel.

5.6.4 Total utilisation of wood components

The Forestry Agency of Japan in 1985 started a four year research project to develop an integrated process for the total utilisation of wood components, in cooperation with 13 private industries. A pilot plant for this process was built in 1989. It consists of the fiberisation of woods, the extraction and separation of wood components, and the conversion of these components into final products (see Fig. 5.16).

The process is based on the "wood explosion" or "steam explosion" treatment of wood. Wood chips are cooked under high pressure at 180-230 °C for 2–20 min, and then suddenly exposed to open air. The product has a popcorn-like quality. The individual components of lignin, cellulose and hemicellulose that are tightly bound together in the wood matrix are forced apart by the steam treatment. Successive extraction with water and 1% aqueous sodium hydroxide solution separates them and they can be purified, decomposed, spun or fermented to yield a variety of products, including xylitol, carboxymethyl cellulose, carbon fiber and alcohol.

The products can be applied widely. For example, the popcorn-like materials can be fed directly to cows, which are ruminants. Xylitol is a sweet carbohydrate that is digested without insulin and can serve as a nutritious drip for people with diabeties. It also retains moisture, and is used as a food additive to fish paste. Xylose, the precursor of xylitol in wood, is not absorbed in the body, and can be used as a dietary sweetener. Lignin is processed by hydrocracking/phenolysis, and is carbonised at 1000 °C to yield

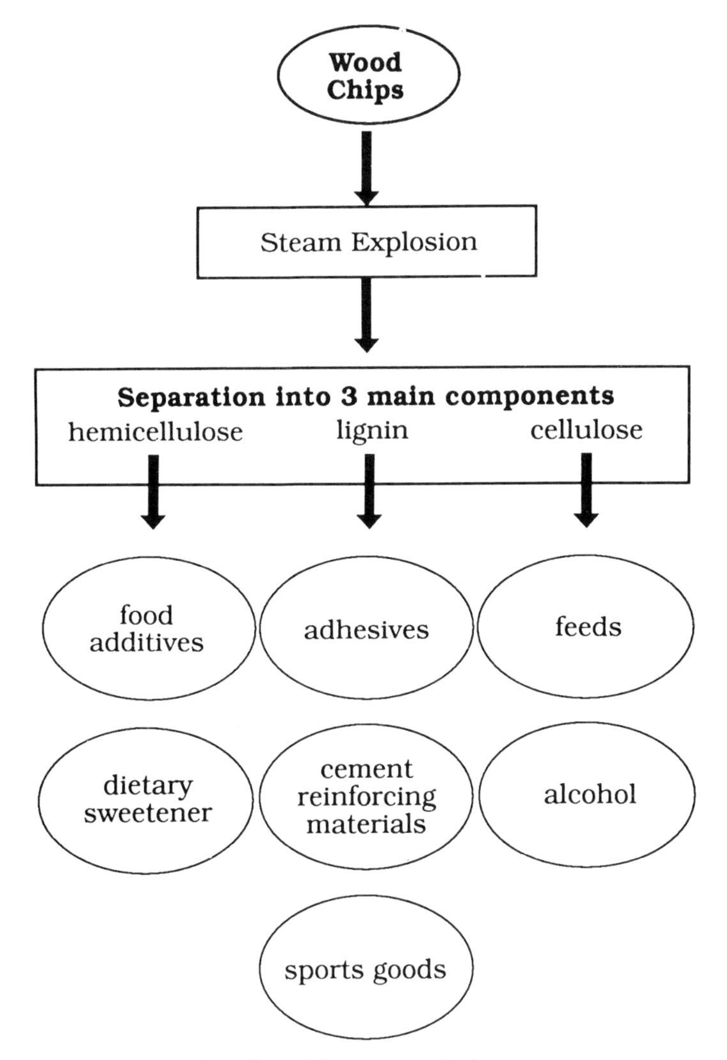

5.16 Total utilisation of wood by steam explosion.

carbon fibers that possess similar physical characteristics to pitch-type carbon fibers.

The Forestry Agency estimate that 30,000 tonnes of wood chips yield 3,500 tonnes of xylitol, 1,800 tonnes of carbon fibers, 6,700 litres of alcohol and 15,300 tonnes of cow feed. The process can be operated commercially, and taps the components of wood not previously usable, namely lignin and hemicellulose. The fibrous cellulosic component, of course, is the raw material for paper production, and for clothing fiber as described in Chapter 8.

5.6.5 Chemical modification of cellulose to provide specificity

Water-soluble cellulose derivatives, such as, for example, hydroxyethyl celluloses (HECs), already find widespread application as thickeners in latex paints, viscosity controllers of mud during oil drilling, foods, cosmetics and building materials. Generally, these have been used in non-specific fashion as bulk commodity materials. Now more detail can be built into the structure to provide greater specificity in function.

A new group of water-soluble polymers with surface-active properties can be prepared by the introduction of hydrophobic substituents, typically C_{12}–C_{18} straight-chain alkyl side chains, into cellulose ethers. Whereas HEC produces viscosity increases in solution through chain entanglement, this new class of polymer has additionally significant intermolecular interactions through the hydrophobic association of neighbouring alkyl side chains. The rheological properties are extremely novel because of the formation of reversible cross-linked polymer networks. Moreover, the solutions are pseudoplastic in behaviour and are considerably more viscous above certain polymer concentrations. Such networks can be broken up by the addition of surfactants or the imparting of stress. Consequently, these properties find specific applications, such as in paint performance and application. For example, a paint of thick consistency flows freely under the shear imposed by the brush, but once on the object again assumes its original high viscosity.

Cellulose sulphate (CS), like the established drug heparin, is an anticoagulant, with its behaviour being closely related to molecular structure. Carboxyethyl carbamoylethyl celluloses (CECECs), in contrast, accelerate the clotting of topical blood and blood fibrin. Various degrees of substitution by carboxyethyl groups or carbamoylethyl groups can be synthesised by reacting alkali cellulose with acrylonitrile. The most effective clotting abilities are determined by a critical distribution of carboxyl and amide groups. The applications in wound treatment will be apparent.

The sodium salt of carboxymethyl cellulose (CMC) prepared from regenerated cellulose, with the crystal form of cellulose II, has a high degree of absorbency towards aqueous liquids. This absorbency of CMC has been harnessed to produce a unique yarn having rubber-like elasticity only under wet conditions. Yarns are substituted to a low degree of substitution (DS) (<0.35) before twisting, so retaining the fiber form. A uniform distribution of the DS over the whole cotton yarn is required to get a fixed shrinkage of the elastic yarn when it absorbs water. A process was developed using a reactor equipped with a liquid circulation so that even contact occurs between the cotton fiber and an aqueous ethanol solution containing alkali and etherifying agent. The yarns are then doubled, twisted and steamed for twist setting. The absorbent elasticated CMC cotton is used in disposable nappies (diapers), and it can replace the conventional elastomer material. Since they show elasticity only when they are wet, they shrink with urine and fit the edge of the nappy to the thighs. Otherwise, the nappy can be quite loose fitting and so retains a comfortable feel, but can still retain the fluid when necessary. It is also much more efficient to manufacture since the yarn is not elasticated during the production process, and it is not bulky or wrinkly.

Cellulose and cellulose acetate have been used for haemodialysis membranes in artificial kidneys because of their good mechanical and permeation properties. The materials, however, are not antithrombogenic, and, therefore, an anticoagulant such as heparin must be administered to the patient to prevent blood coagulation during the dialysis. Now polyion complexes (PICs) have been formed between cationic and anionic cellulose derivatives which produce membranes having excellent blood compatibility *in vitro* and *in vitro*. Moreover, the PICs are biodegradable on implantation. Initially, their mechanical properties were poor. Now the PICs can be localised on the surface of the cellulose membrane so retaining the strength and also imparting blood and biocompatibility.

Such membranes from cellulose sulphate, because of the combination of biocompatibility, rheological behaviour and polyelectrolyte character, are ideal for the encapsulation of sensitive biological substrates. If the substrate is suspended in aqueous cellulose sulphate solution and reacted drop by drop with a suitable cationic polyelectrolyte, a porous membrane of good mechanical quality will incorporate the active biological material and allow controlled release.

5.7 Utilisation of protein functionality

Proteins when present in meat, fish and beans are an indispensable component of our diet. Although there is a wide range of different proteins produced by animals, plants, viruses and other microorganisms, all are made up from only 20 known amino acids. The physical, chemical and biological characteristics of individual proteins are determined by the sequence of amino acids. The amino acids are linked chemically by peptide linkages to form a protein with a particular physiological function. The protein structure is defined conventionally in terms of four quaternary structure (the primary, secondary, tertiary and quarternary structures) according to the particular dimensional viewpoint. Normally a protein has a rather compact high-order quarternary structure, specific to its physiological function. However, the relationship between protein structure and physiological function is still not well understood. The following examples will demonstrate the recent applications of proteins as biomaterials.

5.7.1 Collagen contact lens and slow-release drug systems

Collagen is a protein traditionally used in the food and cosmetic industries. Gelatin used to make edible jelly is made up of collagen, which can also be used as skin protector. Fibrous collagen is the structural material that maintains the organs of the body, and is mechanically as strong as silk. Its physical characteristics make it ideal for use as edible sausage casings.

Collagen also has uses as a biomaterial. When collagen is cross-linked in 5–10% aqueous solution, it forms a transparent hydrogel with high oxygen permeability. This hydrogel can be processed into a soft contact lens, which is increasing its share of the disposable lens market (see Fig. 5.17).

Collagen can also be utilised for drug delivery systems. Often high doses of immunostimulators such as interferon (IFN) need to be taken to maintain a concentration in the blood over required periods. This can induce unfavourable side effects. Such a method of delivery is expensive and ineffective, since most of the administered drug is wasted. The object of a drug delivery system (DDS) is to administer an optimal amount of drug to a specific location over a required period. Pharmaceutical companies are now developing biodegradable types of DDS for slow-release drugs. Sumitomo Pharmaceutical Co. and Koken Co. jointly developed a slow drug-release system using atelocollagen as the drug carrier. Atelocollagen is a three-stranded protein produced from bovine skin by removing the telopeptide regions present at both terminals using the enzyme proteinase. Since collagen without telopeptide regions exhibits extremely low antigenicity, it serves well as a drug carrier.

5.17 Contact lens made from collagen (Koken Co.).

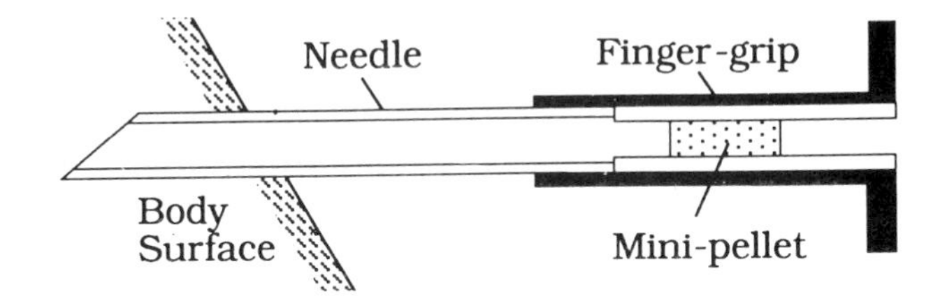

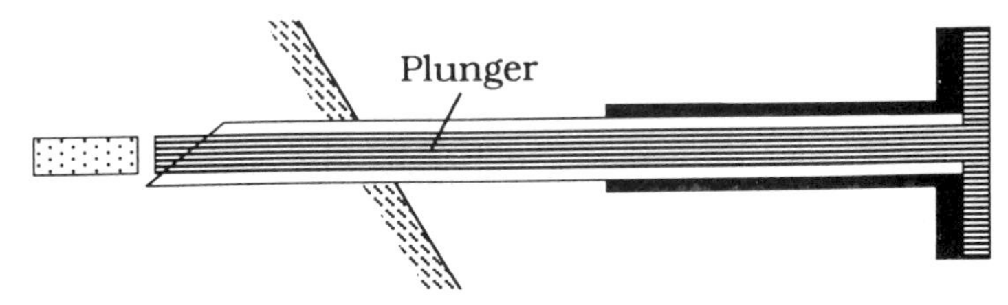

5.18 Device for mini-pellet injection.

The atelocollagen hydrogel containing the drug is injection-moulded into a cylindrical shape, and dehydrated to form a rod-like pellet for injection using a special injector (Fig. 5.18). The drug is immobilised within a collagen cylinder, which swells in the body to release the drug gradually, so maintaining a constant concentration in blood over a long period. Since INF is water-soluble, it mixes well with the hydrophilic atelocollagen. The drug-release duration can be controlled by the rigidity of the atelocollagen gel. It is now being considered for use as an implant to release a drug which can suppress the metastasis of cancer after surgery. Sumitomo Pharmaceutical Co. has carried out a series of animal tests using such an atelocollagen/INF pellet. The successful results were reported first in 1986 at the 13th International Symposium on Controlled Release of Physiologically Active Materials, sponsored by the Control Release Society, USA.

5.7.2 Miracle fruit that changes a sour into a sweet taste

Delicate and complicated tastes are highly sought after in the age of the gourmet. Physiologically, salty, sour, bitter and sweet make up the four basic tastes. In Japan a hot taste is added to these four. Buddhism takes account of five stages in taste development progressing from milk to cheese via yogurt and butter, in accord with the development of Buddha's teaching during his life.

5.19 A red oval "miracle fruit" with 2 cm major axis.

A taste is perceived when the taste cell is stimulated chemically and changes its electric potential to release a neurotransmitter. This results in an impulse in the nerve fiber, which conveys the taste information to the brain. Chemists have been investigating the type of chemical structures which give rise to a particular taste.

A red oval West African fruit, some 2 cm long, is referred to as a miracle fruit (see Fig. 5.19), since it can change a sour taste to make it appear sweet. A sour lemon tastes like a sweet orange after a bite of this fruit. The natives of West Africa bite this fruit to kill a sour taste prior to drinking sour coconut wine or eating sour fermented corn pone. A glycoprotein with a molecular weight of about 25,000 is responsible for removing the sour taste. The miracle fruit flesh contains this particular glycoprotein (miraclin), which interacts with the taste cell and stimulates the sweet taste receptor at its active centre when acid is present. This physiological function could find a variety of applications, particularly in the field of medicine, and now attracts much industrial interest. However, the growth of this tree is too slow for introduction into the Japanese climate. Effort is, therefore, being focused on identifying the active site of the glycoprotein in order to produce a synthetic substitute having the same physiological functions. The development should be valuable for people with diabeties.

Some substances, on the other hand, suppress a sweet taste. If the leaves of *Gymnema sylvestre* (found in India) or *Jububl* tree are chewed, the sweet taste of sugar is eliminated. Gymnemic acid and ziziphin are contained in those leaves, and are responsible for suppressing the sweet taste, without affecting other tastes. Gymnemic acid is a triterpene glycoside made up of gymnemagenin and glucuronic acid. Ziziphin is also a triterpene glycoside, which leads to the formation of the sweet substance, glycylrhizin.

The structure and mechanism of sweet taste receptors are not yet fully understood. Miraclin provokes a sweet taste by interacting with acids. However, in contrast a triterpene glycoside such as gymnemic acid or ziziphin acts only by suppressing the sweet taste, probably by competitively blocking the sweet receptors. Derivatives of gymnemic acid or ziziphin obtained by modifying the side chains can be used to study relationship of structure to the sweet suppression function. Great benefit can be derived from a better understanding between structure and the mechanism of the sweet receptors.

Professor Y. Hiji, from the Medical School of Tottori University, edited the book *Gymnema sylvestre*: *the secret medicine used for over 2000 years in India*. From this book it is evident that sugar taken with *Gymnema sylvestre* leaves is not absorbed in the small intestine. They can, therefore, be used to reduce obesity and control the effects of overeating. It was regarded in India as a wonder medicine, which has been used to treat diabeties for several hundred years.

The structure of sweet substances are well known. If a slight modification is made to the hydrophobic part, the taste changes from sweet to bitter. There is hope, therefore, that the taste of a material can be predicted from its chemical structure, with the possibility that appropriate tastes can be designed into the molecule.

5.7.3 Proteins as bio-stimuli

The life of an organism is influenced by its response to an external stimulus. Although certain non-living substances may also react to the external stimulus, this reaction is more accurately classified as a "change" rather than a "response". Proteins are mostly responsible for the response of living systems.

Monellin is a protein found in the fruit flesh of a tropical plant growing in West Africa, and is 3,000 times sweeter than the same weight of cane sugar. Moreover, its sweet taste lasts longer on the tongue. As noted, miraclin is another glycoprotein that controls taste. There are chemical senses within the body that perceive external chemical substances such as monellin and miraclin, and give rise to the sense of smell and taste.

Insects use pheromones for communication using the sense of smell. Fish perceive chemical substances dissolved in water by smell, as water flows in and out of their noses. A salmon returns to the river of its birth even after four or five years excursion in the sea, following by memory the smell of the river, where it spent its younger days.

Hormones are the chemical substances that produce specific responses. Many possess a peptide structure. R. C. L. Guillemin and A. V. Schally were jointly awarded the Nobel Prize in 1977 for their isolation and identification of the polypeptide hormone (TRH) in the brain. It is a polypeptide hormone of low molecular weight secreted at the hypothalamus which controls the release of a thyroid stimulating hormone.

We still have a great deal to learn about such physiological responses to chemical and biochemical stimuli. It is a field that is being actively researched and will surely improve our understanding and control of body responses.

6 Progression of high-tech fibers

High-functional new materials have been developed by combining the expertise of various industrial disciplines, for example, the materials, information, life sciences and energy industries. Unique and high-value-added products can derive from the synergy between these different industries, as shown in Fig. 6.1. Fiber science and technology is a field that is traditionally strong in adding high value to products by controlling the tertiary structures of the materials, for example, by introducing better orientation or higher crystallinity into the fibers. This technical tradition is now being applied to achieve the higher functionalisation of such materials.

6.1 Utilisation of unused resources

Rayon is produced from alkali-treated wood cellulose, and its fiber can be converted to active carbon fiber. Recently, an investigation was initiated to produce carbon fiber directly from wood as part of a project for the total utilisation of wood.

6.1.1 Carbon fiber from wood

The wood industry is a typical biomass industry, and a variety of research projects is underway to convert wood into more valuable products.

Carbon fiber is produced mostly from PAN at present. Its study was started 30 years ago when it was produced by carbonising rayon derived from cellulose. At one time, lignin, a wood component, was used to produce carbon fiber by mixing it with polyvinylalcohol and combusting. Thus it was natural for the Oji Paper Co. Ltd to consider producing carbon fiber directly from the wood. Wood is composed of several components. The carbonising conditions must, therefore, be carefully controlled to take account of the physical characteristics of the components since each has its own functional qualities.

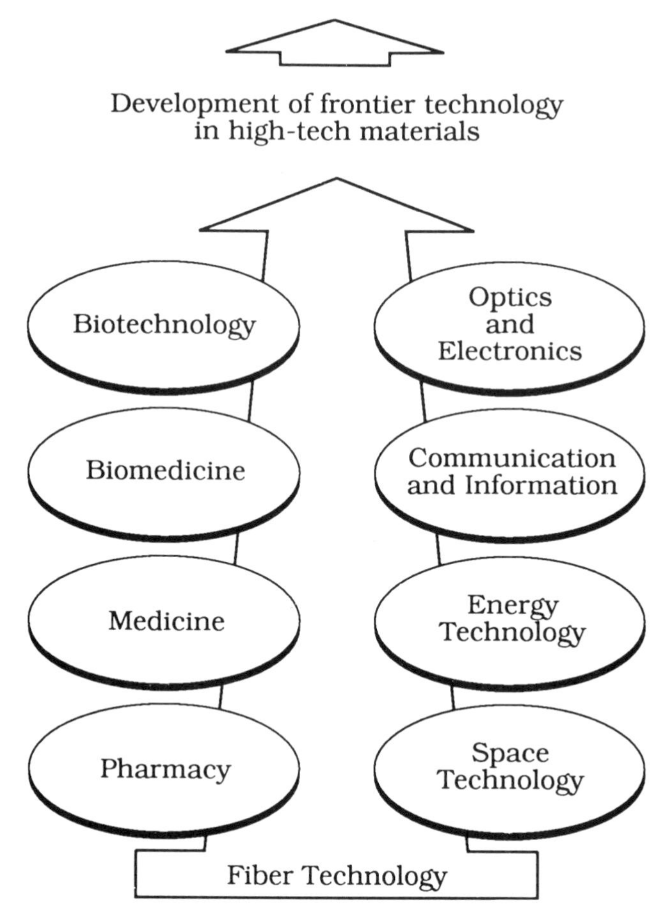

6.1 Synergies between different industries which can contribute towards providing new materials for the 21st century.

For example, cellulose is characterised by its elongational fibrous property, whereas lignin is responsible for the high carbonisation yield. The different functional groups of wood components control the heterogeneous surface structure of the resulting carbon fiber, which would, therefore, have quite different characteristics from that derived from a single component material, such as PAN or pitch. Most of the attempts to produce carbon fiber from lignin or from various mixed compositions of wood components failed. The appearance of the product was not satisfactory from these attempts.

The Oji Paper Co. found that it was possible to spin wood directly if it was dissolved, in a suitable solvent. They examined the best spinning conditions of the acetylated wood dissolved, for example, in phenol, in cooperation with Kyoto University. A cross-linking agent added to the acetylated wood solution improved its spinning characteristics. The resulting fiber is thermoset for a short time at 250 °C, and then carbonised to yield carbon fiber of better quality than the pitch-type, particularly for general uses. The physical characteristics are somewhat similar to that of Kynol fiber. However, its high production cost, because of the low carbonisation yield, prevents the use of the method on a commercial scale at present.

6.1.2 Fibrous active carbon derived from cellulosics

Toyobo Co. has succeeded in achieving the commercialisation of fibrous active carbon produced from cellulosics. The process incudes both carbonisation and activation. Since cellulosics are flammable, they must first be processed with a flame-retardant, and burnt at a relatively low temperature (200–300 °C) to decompose the organic components into carbon and produce a carbon fiber with sufficient strength for practical applications. The carbon fiber is then activated by treating with carbon dioxide and steam to form fine pores. The temperature and duration of this activation process is critical to produce good fibrous active carbon.

Fibrous active carbon now finds enhanced utilisation as the social demands for pollution prevention and energy conservation increase. The starting material, the cellulosic source, is coconut husks. Compared with conventional granular active carbon, fibrous active carbon exhibits a better adsorption quality owing to its large contact area with air or liquids. The pore size at its surface is approximately 10 Å in diameter, so each pore traps a molecule. The adsorption rate depends on pore size, its position and distribution at the surface. The specific pore surface area is some 1.5–5 times larger than that of granular carbons. Fibrous active carbon can be woven, knitted or converted into a paper sheet. The corrugated cardboard form of fibrous active carbon is now utilised as the base material for the solvent adsorption/desorption devices used in car factories. It is capable of adsorbing various solvents present at low concentrations in exhaust fumes, or arising from the painting process. The system can be regenerated by blowing heated air over the active carbon, which desorbs the material.

Biotechnology has been used for generations in the fermentation processes to produce wine, beer, sake, etc. Alcohol is inevitably released into the air during fermentation, and major brewers are now investigating the recovery of the released alcohol using an adsorption device made from active carbon cardboard. It can then be further used. The released gas during fermentation contains a variety of components. Some, particularly low molecular weight components, such as acetaldehyde and ethyl acetate, produce a foul smell. These noxious components can permeate fibrous absorbents. Only components in the higher molecular weight range can be recovered with this adsorption device. Nevertheless, their recovery can be most useful. Following the fermentation of soya sauce, the recovered liquid is returned to the stock solutions, contributing a considerable resource-saving.

Powdery or granular active carbons of the conventional type have also been used as catalyst carriers, for solvent recovery, the decoloration of sake and sugar, etc. within the chemical and food industries. In view of the new developments, many more applications can be envisaged.

6.2 Biotechnology and fibers

Biotechnology has developed rapidly since 1953 when the gene recombination technique was first established. Now medical/pharmaceutical, foods, fermentation and chemical industries make extensive use of gene recombination, cell fusion, mass cultivation of animal/plant cells and bioreactors.

Now, major fiber-related industries are actively introducing biotechnology, and have invaded the bio-related fields of medical/pharmaceutical, fermentation and foods industries, as summarised in Table 6.1. Japanese cooperation with foreign industries and research institutes has been expanded in the biotechnology field to counter the extreme competition in this field. We should soon see biotechnology widely expanded in the fields of fermentation, foods, resources, energy and agriculture industries. As illustrated in Fig. 6.2, the fiber-related industries are using biotechnology for (i) making changes to production processes; (ii) utilising fibers as biotechnology materials; and (iii) for producing new useful materials.

As an example of (i), biotechnology is now being applied to breed a new type of cotton. Biotechnology has also made possible more efficient energy-saving processes for fiber production and the production of fibers of higher quality and higher functionality. In relation to (ii), fibers are already being used for the making of artificial internal organs and other medical materials. Further applications are imminent in the field of biosensors and bio-electronics. Area (iii) relates to the use of fibrous materials such as fibers, pulp and wood wastes obtained as biomass for producing useful materials by enzymic and microbiological action.

Here we describe the work of Professors Y. Ikada and T. Hayashi of the Research Centre for Medical Polymers and Biomaterials, Kyoto University, in producing medical materials via bioreactors.

Table 6.1. Fibers in biotechnology

Realise a plan	Hardware (raw materials)	Software (examples of utilisation)
	Hollow fiber	Artificial kidney, artificial lung, water deionisation, condensation of alcohol, oxygenator
Short to medium	Antithrombotic or haemolytic fiber	Blood cell exchange, plasmaphoresis
	Carrier fibers for biocatalyst	Drugs, foods, microanalysis
	Resolving fiber	Agricultural and gardening uses
	Slow-release fiber	Drugs, agricultural chemicals
	Biodegradable fibers	Sutures, artificial blood vessels
Medium to long	High-function membrane	High-performance biosensor
	Bacterial fiber	Fuels, chemicals, chemical manure, feed, physiological active materials
	Biomedical material	Artificial heart, artificial muscle, hybrid artificial liver
	Bioelectronics	Biosensor, artificial cell membrane, biocomputer

6.2.1 Fibers for medical uses

There are some 8 million users of contact lenses in Japan, making them the most extensively used items of medical equipment, followed by the artificial kidney where hollow fibers are used for the haemodialysis membrane. There are other major medical use of fibers as shown in Table 6.2. Filaments or threads are used for suture and absorbents (artificial liver). Hollow fibers are applied in blood purification, the artificial kidney and artificial lung. Although small in quantity, knitted or woven fabrics are used for artificial blood vessels. Cotton, non-woven fabrics and gauze are the traditional medical materials for haemostatic dressings and for artificial skin. Table 6.3 summarises such biomedical uses of fiber materials.

6.2.1.1 Sutures

Two types of sutures are currently available: (i) the assimilated type, such as a catgut; (ii) the non-assimilated type such as silk or polyester filament. Inevitably there will be a move to make more use of the readily resorbable type and a greater utilisation of polyester or polypropylene filaments for external use. Catgut, the most widely used assimilated type, is made from collagen, extracted from ox bones. However, it is not particularly suitable for implantation for biocompatible reasons. The human body normally rejects foreign materials, and collagen is not an exception when implanted directly

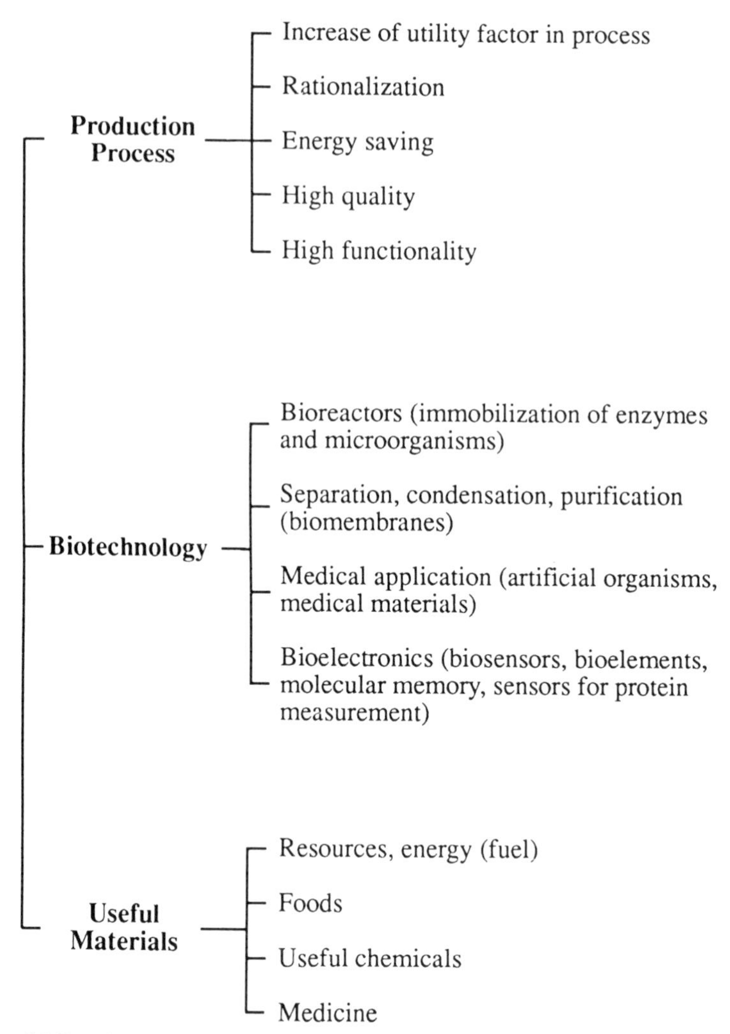

6.2 Development of biotechnology-related fibers.

into the body. Another weakness of the catgut suture is its strength, which deteriorates by a half after a week in the body, despite the fact that three weeks are required for the recovery of an incision after surgery. Thus active investigations are underway to produce a biocompatible suture of the assimilable type made from synthetic polymers.

A mono-filament is preferable for use as a suture because of its smooth surface. Poly(glycolic acid) is a synthetic polymer currently used for sutures of the assimilable type, but it is too hard a material to make a mono-filament suture. Thus it is now used to make the multifilament or blended type of

Table 6.2. Production of artificial organs

Articles		Total cost (million US$)	Quantity
1	Mechanical lung and heart	24	94,682
2	Pacemaker for heart	61	60,561
3	Artificial valve for heart	18	5,359
4	Haemodialysis	45	7,850
5	Dialyser	344	12,217,250
	(hollow fiber type)	(305)	(11,089,142)
6	Circuit for blood	64	12,596,283
	(for haemodialysis)	(46)	(10,866,672)
7	Instrument to purify blood	11	229,250
8	Artificial pancreas	1	11
9	Artificial blood vessels	6	13,306
10	Artificial joint, artificial bone	38	46,134
11	Suture	27	56,215,209
	(non-absorbable)	(17)	(48,938,311)
	Total	**639**	

suture, and this is commercially available now. To overcome its hardness, and rough surface, it is normally coated with a plasticiser, and in consequence the suture must be knotted several times to ensure that it does not loosen.

The ideal suture is a mono-filament with a smooth surface that can pass through the skin without being caught, and can be tightened with a single knot. The polybutylene teraphthalate (PBT) mono-filament suture is currently the most popular because of its acceptable strength and smooth surface. The poly(glycolic acid) suture is used currently for heart surgery in order to withstand the high pressure within the heart. However, poly(glycolic acid) is not fully assimilable by the body for one to two years. The objective is to reduce this time to less than six months. Johnson and Johnson Inc. have developed the suture with trade name Maxon, which appears to satisfy all the required criteria.

6.2.1.2 Blood purification

Table 6.4 identifies the three major artificial organs related to blood purification. Various fiber materials are now used in the manufacture of these organs.

Artificial kidney

Haemodialysis is indispensable for people suffering from kidney disease, and some have been subjected to this treatment for over 25 years. A complication is that long-term use of the haemodialyser (artificial kidney) can introduce another problem, because some unwanted substances in the blood are not

Table 6.3. Fibers for biomedical uses

Medical use	Examples of fiber materials	Key points for medical use
Fiber materials in hospitals	Cellulose, nylon, polyester	Sterilisable, dust free, easy handling
Suture	Catgut, silk, cotton, polyglycolic acid, polylactic acid, polydioxanone nylon, polypropylene, polyethylene, polyethylene terephthalate	Sterilisable, knotting strength, tissue compatibility, absorbable or non-absorbable, softness, easy handling
Surgical tape, tissue adhesives	Collagen, oxycellulose, fibrin, polyurethane	Ease of preparation, biocompatibility, resorbable, sterilisable
Vascular implant	Dacron, Teflon, biograft, collagen, segmented polyurethane, polyglycolic acid, polyester, urethane	Blood compatibility, porous structure, resorbable, easy for tissue ingrowth, minimise clotting
Artificial skin	Natural skin equivalent (dried pig skin, collagen, chitin), silicon/nylon, polypeptides, silicone/collagen/Glycosamino glycans, hybrid skin equivalents	Adhesive to tissue surface, prevent loss of fluids and infection, resorbable, relief of pain
Soft tissue implant	Silicone, polylactic acid, collagen	Relief of pain, adhesive, resorbable
Joint replacement	Polymethylmethacrylate, high molecular polyethylene, silicone	Mechanical strength, wear durability, must not become loose
Artificial bone	Polyethylene/apatite, polylactide/apatite, polysulphone, carbon fiber, polyethylene terephthalate, glass ceramic	Moderate mechanical strength, adhesion to bone, resorbable
Artificial kidney	Cellulose, polymethylmethacrylate, copolyethylene-vinyl alcohol, polyacrylonitrile, polysulfone, polycarbonate, chitin, chitosan	Moderate mechanical strength and permeability, blood compatibility, suppression of complementary activation
Artificial lung	Silicone, polypropylene, polysulfone polyethylene, Teflon	Gas exchange effect, blood compatibility, suppression of blood plasma leak
Artificial liver	Carbon fiber, cellulose, polyetherurethane, poly(HEMA)	Blood compatibility, adsorptive activity
Carrier for DDS	Polysaccharide, collagen, nylon, polyacrylonitrile, polyvinyl alcohol	Large surface area, porosity, functional group

Table 6.4. Blood purification devices

	Artificial kidney	Artificial liver	Artificial lung
Function	Excretion of waste materials and water. Control of electrolyte concentration	Separation and disposal of patient's plasma and supply the fresh plasma into the patient's blood	Execution of the cardiopulmonary function of the patient
System	Haemodialysis (HD) Haemofiltration (HF) Haemodiafiltration (HDF)	Plasmapheresis (PP) Haemoperfusion (HP)	Haemoperfusion (HP)
Material	Cellulose hollow fibers Polyacrylonitrile Polymethylmethacrylate EVAL	Cellulose hollow fibers Anionic exchange resins Active carbon powder Ceramics	Polypropylene, silicone Silicone hollow fiber Silicone–acrylamide grafts

dialysed out. Also, an immunological reaction can occur with repeated use, since the dialyser is a foreign substance to the human body. When the blood is circulated through the dialyser, the leucocyte count in the blood decreases over the first 20 min of dialysis, but recovers to its original value after 1 h. The haemodialyser is made up from a bundle of hollow fibers through which the blood circulates. The objective is to improve the surface of hollow fibers so that the leucocyte decrease does not occur. Although some materials do not appear to be dialysed from the blood, certain cases of uraemia and other kidney troubles are nevertheless improved on dialysis. There is no definitive explanation for this improvement at present, although there are some hypotheses. Kidney troubles, it is believed, can be caused by proteins of molecular weight between 10,000 to 30,000. The blood also contains substances that must be retained, such as albumin with a molecular weight of about 70,000. Each hollow fiber manufacturer is now developing a suitable membrane, through which the harmful proteins of molecular weight around 20,000 pass but not the proteins of the molecular weight around 70,000. These could have much practical benefit.

Mechanical lung

Silicone or polypropylene hollow fibers are used for the fabrication of the mechanical lung to allow permeation of gases. It ideally should function for at least one to three weeks. However, the present mechanical lung lasts at most one week, because its ability to remove carbon dioxide falls off. The lung is a form of gas exchanger to supply oxygen to the blood and remove carbon dioxide. The mechanical lung was first developed as a device to replace lung function during heart surgery, and is now extensively used for this purpose in the USA (about 250,000 per year) and Japan (20,000 per year). A newer form of mechanical lung can also be used as a supplementary respiratory device over a longer term to assist the breathing of patients suffering from acute lung or heart failure, or older people with weak lung function.

The artificial lung of the micropore membrane type is the presently used system in which oxygen comes into contact with blood via the membrane in the same way as in a natural lung. Mitsubishi Rayon Co. has developed a microporous polypropylene (PP) hollow fiber for the manufacture of an artificial lung (see Fig. 6.3), and is currently supplying the fiber to medical device manufacturers. Here gases freely pass through the pores of PP hollow fibers, but not the blood, because of the hydrophobicity of PP membrane. As a result, the artificial lung of the gas-bubble type is rapidly being replaced with the membrane type, which could soon dominate the artificial lung market, according to the estimates of the Mitsubishi Rayon Co. PP hollow fiber exhibits good compatibility with blood and excellent gas permeability. Its use allows the design of a compact artificial lung that is easy and safe to operate. However, its long-term use causes a leak of blood plasma components, and an investigation is underway to improve the membrane material in order to eliminate this disadvantage.

Artificial blood vessels

Artificial blood vessels are now commercially available, and are made mostly from polyester or Teflon. They are used to replace thick arteries or veins of 6 mm, 8 mm or 1 cm in diameter. Although polyester is biocompatible, its anticoagulant activity is poor. Porous Teflon, on the other hand, exhibits both good biocompatibility and anticoagulant activity. However, thin blood vessels (diameter less than 3 mm) made from Teflon lead to other problems. Consequently, at the present time, the coronary artery or the thin veins in hands or legs are replaced with blood vessels from other parts of the same human body. Thus, research in this field is targeted to produce thin artificial blood vessels of diameter less than 3 mm.

Dr M. Kodama, Research Institute for Polymers and Textiles, Japan, has developed an artificial blood vessel of inner diameter 1.5 mm, which has a three-layered structure made up of collagen, heparin and Teflon. The inner layer of collagen and heparin imparts good anticoagulant activity, whereas the porous Teflon tube which makes up the middle layer provides the mechanical strength. It is coated with collagen to form an outer biocompatible layer.

The task of producing an artificial substitute for the bile duct can be approached in a similar manner to blood vessels. However, the bile duct exhibits two contradictory physical properties, in being extremely soft but also very tough. The bile duct cannot be cut using a thread, whereas a soft synthetic polymer tube slices easily this way. A synthetic polymer material

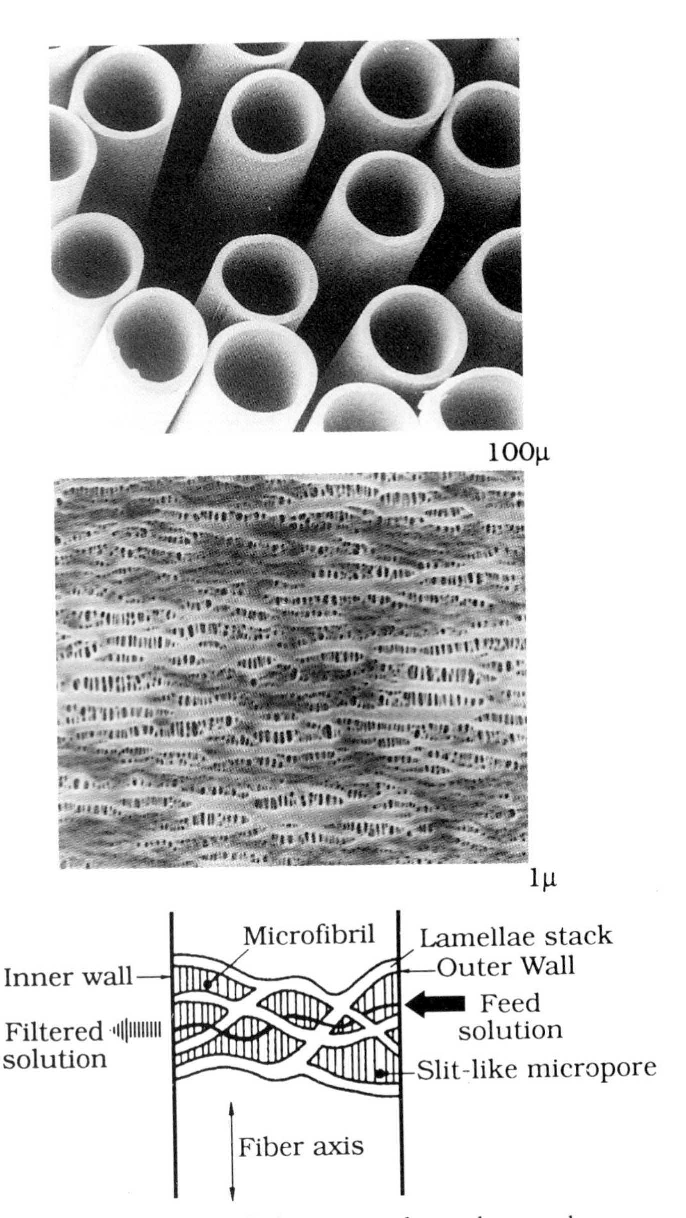

6.3 Schematic structure of microporous polypropylene membrane (longitudinal cross-section).

which resists thread tightening, is generally too hard and causes other problems during surgery. The race is now on to develop a soft but tough synthetic polymer material suitable both for artificial blood vessels and bile ducts.

6.2.2 Bioreactors

Biotechnology is often regarded as one of the most advanced modern technologies. Yet fermentation technology is traditional in Japan, and much recent biotechnology has been developed on this traditional basis. Newer developments use more recent technologies, such as gene recombination, cell fusion, large-scale cell cultivation and bioreactors.

The bioreactor utilises the specific function of biocatalysts to produce new and useful fine chemical materials effectively, to generate energy, for use in quantitative and qualitative analysis and to remove pollutants. In effect, it is a type of immobilised biocatalyst, utilising enzymes. Over 2,000 enzymes have been found suitable for use in a bioreactor, but few are in commercial use at present. If, however, organelles, microorganisms or cells could be immobilised, the bioreactor's efficiency would increase sufficiently to allow practical use.

Enzymic reactions proceed under extremely mild conditions compared with the usual type of chemical reaction. Thus effective use of enzymes in chemical processes leads to energy conservation. They are also characterised by substrate specificity; the reaction yield is high, and there is no by-product, so the process can be economically favourable.

How then can fibers be applied as carriers for immobilisation processes and what are the advantages? Bead-like, spherical or membrane-type carriers are conventionally used to immobilise biocatalysts. However, a fibrous carrier is capable of providing an infinitely extended surface area. Porous, non-circular cross-sections, fine and/or hollow fibers can offer a large specific surface area without break. Such a fibrous carrier can be processed in the form of a thread, braid, cloth, net or a non-woven fabric, according to its applicational environment. Porous fibers can be made in the form of filter paper or non-woven fabric and used as a biosensor. With recent technology, pores of homogeneous size can be produced on the fiber surface to entrap relatively large substances such as enzymes (over 4 μm) or bacteria (a few tens of μm).

Various methods are currently employed to immobilise biocatalysts. The most popular is to wet-spin the mixture of biocatalyst and carrier in order to entrap the enzyme or bacteria within the fiber. Many pores can be formed on the fiber surface by wet-spinning in such a way that substrates and reaction products freely pass through, but biocatalysts of relatively large size remain entrapped. Biocatalysts can be immobilised by simple adsorption on to the porous fiber surface. If biocatalysts need to be fixed more firmly, they can be covalently or ionically bound to the fiber. For example, penicillinacylase can be immobilised on a reduced porous polyacrylonitrile fiber by covalent linkage to produce 6-amino penicillin, which serves as a raw material for reactive penicillin. Several thousand tonnes of 6-amino penicillin are produced annually in this way. Enzyme-immobilisation by covalent linkage is currently a most active research field.

With the advent of more old people in society, clinical analysis will increase in importance as a support for preventive medicine. Efficient biosensors need to be developed to ensure rapid, economic and effective analysis. Biosensors using immobilised biocatalysts of the covalent linkage type are very suitable for quantitative analysis, since no biocatalyst is lost after the reaction. Moreover, this type has another advantage because the biocatalyst is immobilised on the carrier surface and can contact the substrate without hindrance. Such a device is now being utilised in the food industry with amylase, in the pharmaceutical industry with penicillinamylase, and in the medical industry with urease (to detect urea in blood).

In some instances, enzyme-immobilisation by adsorption is favoured. Since the enzyme here is loosely linked on the fiber surface, its activity is high but it is desorbed easily. Consequently, the carrier can be regenerated by replacing the original enzyme with a replacement, after the enzyme has lost its activity after repeated use. This type of the bioreactor is used in acetic acid production, by utilising acetobacter adsorbed on to cotton-like polypropylene fiber.

Bioreactor design is also important if an immobilised enzyme is to be harnessed effectively. The choice depends on the application. There are many types available, for example, the integrated type which mixes the substrate and immobilised enzyme, or the flux-flow type, where the substrate flows through uniformly packed beads of immobilised enzyme. The beads can be replaced with a bundle of hollow fibers with the immobilised enzyme on the surface, and the substrate is transported either in the fiber hollow or along the outer surface of the fibers.

Although bioreactors are generally recognised as extremely useful, their practical application has not yet been fully realised.

6.2.3 Magic fiber for AIDS diagnosis and treatment

The Asahi Chemical Industry Co. has developed a porous hollow fiber membrane Bemberg Microporous Membrane (BMM) to filter out and isolate the AIDS virus and hepatitis type B in blood. BMM is made from cellulose fiber (Bemberg) regenerated from cuprammonium solutions of cotton linters.

Synthetic polymers are known to cause blood clotting as a result of protein adsorption. However, regenerated cellulose is free from this problem, and for this reason is used for the artificial kidney in the form of hollow fiber. In order to allow proteins to permeate, but to isolate viruses using the same membrane, it is necessary to have homogeneous pores in the membrane that are larger than proteins but smaller than viruses. The Asahi Chemical Industry Co. has established the technology to produce such cellulose membranes having homogeneously distributed pores of predetermined diameter. Spherical B-type hepatitis virus and the AIDS virus have a diameter of 42 nm and 90–100 nm, respectively. Thus the membrane needs to have pores of 30–40 or 40–75 nm in diameter, respectively, to isolate these viruses. A single layer of membrane is not sufficient to isolate such viruses completely; consequently BMM has a multi-layer structure of 100–150 layers. This multi-layer hollow fiber membrane is produced by wet-spinning from cuprammonium solution of cotton linter mixed with an organic solvent. The solution undergoes phase separation and is composed of two phase made up of a concentrated and a dilute organic solvent. The concentrated phase forms a continuous organic solvent layer, and the dilute phase is made up of small organic solvent holes of a uniform size in the cotton linter solution. When spun, the resulting hollow fiber is made of 100–150 layers of cellulose membrane, with pores of a predetermined diameter (see Fig. 6.4). The pore size and the degree of crystallinity of BMM depends on many external factors such as temperature, solvent composition, component purity and time. Usually BMM is 300–400 μm in outer diameter, 250–350 μm in inner diameter, and is composed of 100–150 layers of membrane of 25–40 μm in thickness. The actual module is made of 300 BMM hollow fibers which together are 3 cm in diameter and 15 cm in length. Each layer of BMM has over a billion pores which enables complete filtration and isolation of the viruses.

Yamaguchi University and the Hokkaido Red Cross Blood Centre have cooperated with the Asahi Chemical Industry Co. to use BMM for the isolation of AIDS and hepatitis B virus. They first reported the work at Third International AIDS Conference held in Washington, USA, in 1987. Professor N. Yamamoto, School of Medicine, Yamaguchi University, demonstrated that he was able to isolate the AIDS virus completely from a culture solution of human AIDS virus.

BMM is now expected to be applied to the diagnosis and treatment of AIDS. At present, AIDS is diagnosed by using the antigen–antibody reaction, but using this method the result is negative until an antigen is formed. Since BMM isolates the AIDS virus itself, AIDS diagnosis by BMM is possible even in the early stage of virus infection, even before the antigen is formed. Although BMM does not influence directly the AIDS virus, it is capable of removing virus from plasma and so suppresses its multiplication. AIDS virus immersed into lymphocytes grows there, and then overflows into plasma. If the isolation rate of virus from plasma is fast, the clinical progress of AIDS can be suppressed. This suppression of the AIDS virus can allow the

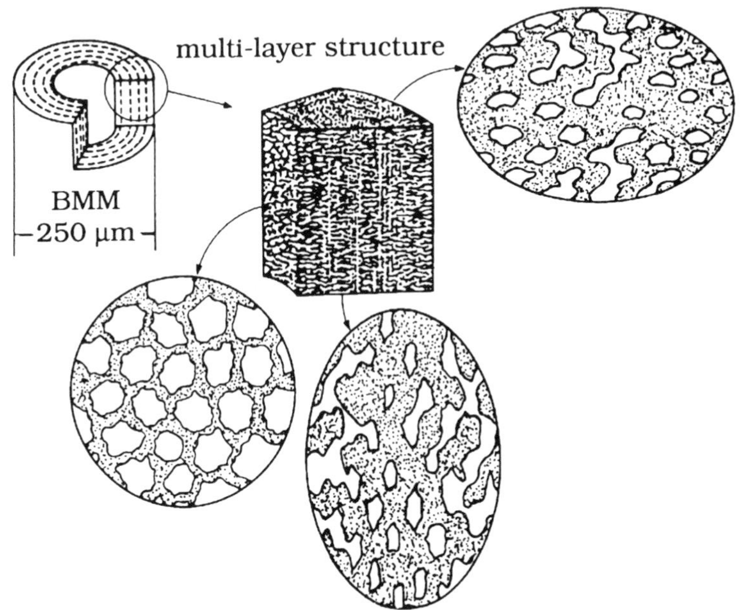

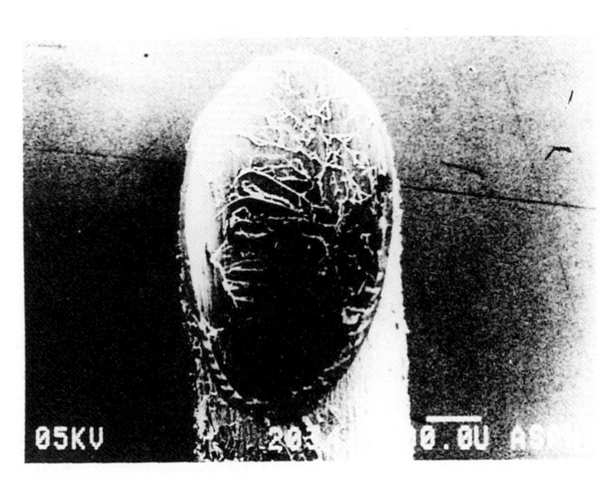

6.4 BMM is made from cellulose fiber (Bemberg) (Asahi Chemical Industry Co.).

reactivation of the metabolic functions of the human body, so that treatment efficiency will improve when combined with other medical treatments.

Other applications of BMM are found, for example, in the complete isolation of virus during plasma medicines manufacture, the administration of fractionated plasma-producing medicines for haemophiliacs, and the prevention of virus infection during ordinal plasma transfusion. BMM is also useful for the isolation of hepatitis non-A non-B virus and in the study of unknown viruses or other physiologically active substances.

6.3 Electronics and fibers

Clean rooms are indispensable for the production of VLSI circuits, which herald the sub-micrometre age in the fields of semiconductors and biotechnology. The degree of required cleanliness is continually increasing as the scale of integration and precision in the VLSI industry improves. The demand for cleanliness is increasing also in other production areas, such as that of magnetic discs and cathode-ray tubes. Indeed, no high technology can develop to its extreme without even stricter cleanliness.

The production yield in a clean room falls in to the presence of dust, which mostly (40–50%) comes from workers in the room. Thus synthetic fiber manufacturers are working energetically on the development of dust-free garments of high performance to meet this demand, by improving the dust-proof and electrostatic-proof characteristics of the materials. Functionalities such as safety, thermal insulation, wear comfort and working efficiency are primary requisites for working clothes. Although the design of working clothes for the clean room is constrained by the specific environment involved, a fashionable element needs to be introduced to add mental comfort to the functionalities required. Working clothes must be free from electric charges when used in a clean room for the production of semiconductor devices to avoid static electricity, but should also be durable against repeated washing and steaming necessary for sterilisation against bacterial contamination.

The main criteria for working clothes suitable for the clean room are; (i) the dust generated from underwear must not leak through the working clothes; (ii) no dust must escape from the clothing material itself; (iii) the clothes must be antistatic and dust-proof; (iv) they should be moisture-permeable and comfortable; and (v) the clothing should be durable against washing (see Table 6.5).

Polyester is probably the best synthetic material currently available for clean room working clothes. Carbon is mixed into the polyester resin to make it electrically conductive. These conducting fibers can be woven into high-density fabrics and eliminate static electricity induced by the low humidity atmosphere of the clean room. Aramid fiber is used when high fire-resistance is required. Natural fibers such as cotton or wool are not applicable for this purpose because of the dust they generate.

The degree of cleanliness is evaluated in terms of the number of dust particles per unit volume atmosphere. The highest degree is denoted as Class 1, where a unit cubic foot (0.028 m^3) of air contains less than one dust particle

Table 6.5. Properties required for garments for use in clean rooms

Required property	Contents	Remarks
Dust-proof	Garments should not cause dust. Ravellings and processing agents must not be released	Dust should not come out from inside of garment, especially from nape and cuff
	Cloth for clean garments should have ability to filter dust from human body and underwear	
	Surface of clean garment should not adsorb dust	
Charge-proof	Garments should not adsorb dust	Destruction of semiconductor and explosion of flammable gas should not be caused by spark discharge due to garments
	Cloth for clean room must not induce static electricity	
Durability	Garments should be easy to wash and no need for ironing	Durability for 100 washings is necessary
	Garments should have durability to washing, steam sterilisation	Water and tetrachloroethylene are used for washings
	Garments should be durable to chemicals	Steam sterilisations are carried out at 120–130 °C for 20–30 min
Wearability	Easy to work in the garments	The garment design must allow ease of work
	Cloth is not stuffy and has good handling	
	Cloth is not "see-through"	
Others	Cloth is suitable for sewing	

of maximum 5 μm in diameter. The degree of cleanliness is graded down as Class 10 (less than 10 dust particles), Class 100 (less than 100 dust particles), Class 1000 (less than 1000 dust particles), and so on. A clean room meeting Class 1 to Class 10 criteria can only be achieved by strict control of the air conditioning and filtration systems.

The total demand for clean room working clothes was some 1.6 million items in 1996 in Japan, with 54% used in the electronics industry, 13% in the pharmaceutical industry, 11% in the food industry, 9% in the precision engineering industry and 13.7% in biotechnology, medical research and printing. Although the required degree of cleanliness depends on the type of industry, 66% require a degree between Class 10,000 and Class 1,000, 20% Class 100,000, 12.5% Class 100 and 1.5% Class 10. Each year a higher degree

Table 6.6. Development of clean rooms for industry with the required degree of cleanliness

	(1970)	→	(1980)	→	(1985)	Frontier at present	
Integrated circuit	1 kbit	4 kbit	16 kbit	64 kbit	256 kbit	1 Mbit	4 Mbit
Minimum pattern dimension (μm)	8	5	3	2		1	
Diameter of particle to be removed (μm)	1	0.8	0.5	0.3	0.2		0.1
Degree of cleanliness (particle/ft^3)	10,000	1,000		100		10	1
				←	ultra-clean room		→

of cleanliness is required by the semiconductor industry. Classes 1,000 to 10,000 were sufficient a few years ago, but now Classes 0 to 1 are called for to accommodate the increase in IC capacity of semiconductors. IC capacity is expressed in terms of bits. A few years ago, 256 kbit was the most integrated circuit available, but now a 1 Mbit IC is commonly used (see Table 6.6). The line width of the 1 Mbit IC pattern is 1 μm, so that a dust particle of 1 μm damages the pattern. The maximum dust size allowed in a clean room is 1/10th of the pattern line width. The IC capacity is now increasing from 4 Mbit to 16 Mbit, and as a consequence Class 0 or Class 1 is required for the degree of cleanliness in the production of these high-density semiconductors. Humans must be regarded as a major dust source, so that it is unlikely that any human can enter a clean room of Class 0 or Class 1. A robot replaces a human worker in this situation. Even so, a robot also generates dust due to abrasion between its moving parts, and consequently this too must be reduced to a minimum.

Humans foul clean room air by dropping dandruff, old skin and dust from underwear. Each apparel maker is now developing dust-free working clothes which envelop a person completely and ensure that there is no leak of dust from the seam or neck areas. For example, Toray, in cooperation with Professor T. Ohnishi of Tohoku University, developed the Toray-type dust-free working clothing. The neck is made of Spandex, the edge of the sleeve is double to prevent dust release, the fastener is positioned at the side, and it is fitted with an integrated cap having a knitted interior to envelop the hair.

Kanebo cooperated with Shimizu Construction Co. and Sharp Corporation to develop the sucking-type of dust-free working clothing. This system is fitted with a small electric vacuum cleaner (weighing 1 kg) attached to the waist with a sucking vest which filters the air inside the clothing and exhausts clean air into the atmosphere.

The Asahi Chemical Industry Co. is now selling the M-bit clean suit

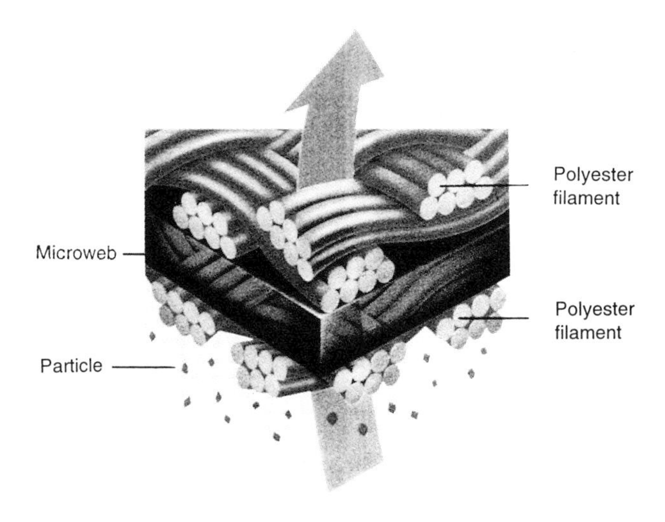

6.5 Structure of Hepa Cloth (Asahi Chemical Industry Co.).

Hepawear. This is dust-proof working clothing made from Hepa Cloth (see Fig. 6.5), which was developed as Hepa (high-efficiency particulate air filter) in the USA for military and high-technology purposes. Hepa Cloth is able to filter the air and collect fine particles of 0.3–0.5 μm in diameter. The material has a three-layered structure composed of a dust-proof non-woven fabric, made from a fine polyester filament of 1–2 μm in diameter, sandwiched with polyester clothing material. It has a dust filtration capacity of 0.3 μm (according to JIS Z-8901). Such Hepa Cloth has a low pressure loss, good dust-proof ability (over 80%), and air permeability of about 10 cm^3/cm^2 s. Hepawear partly employs Hepa Cloth, and is designed to ensure free body movement, comfort and good moisture permeability.

Teijin has developed the antistatic and dust-proof clothing Selguard C, which is a two-layer polyester fabric laminated with a moisture-permeable antistatic carbon-containing film 25–30 μm thick. Up to 2000 V is allowed for static electricity generated during wearing dust-proof clothes, but Selguard C generates less than 50 V static electricity. It is, therefore, ideal for use in the production of semiconductors and prevents the damage caused by electric discharge. For this purpose 0.6 μq per suit is acceptable in a clean room working to JIS standards, but it is often necessary to reduce below 0.3 μq. Selguard C achieves less than 0.1 μq in terms of the charge generated per suit.

Elitron, developed by Toyobo (see Fig. 6.6), is an electrostatic air filter made of electret fibers having a high dust filtration efficiency and low pressure loss. These electret fibers retain positive and negative charges, and collect dust

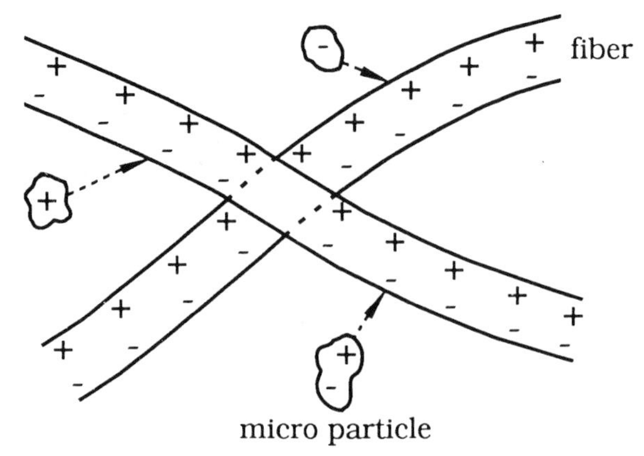

6.6 Mechanism for the gathering of particles by Elitron (Toyobo Co.).

particles from the air by electrostatic forces. Elitron is now used to produce air cleaners and air conditioners for cars, and also for various types of filters, including dust-proof masks.

The Mitsubishi Electric Corp. described the necessary requirements for dust-proof clothing (Fig. 6.7). These are (i) a double enveloped head with a two-layer structure comprising a net cap and hood, leaving no space between the hood and the head; (ii) a double protective edge for the sleeves; (iii) a zip-running fastener; (iv) an air path at the waist; and (v) incorporating low-dust-producing underwear made from polyester.

6.4 Cars and fibers

The trade friction between the USA and Japan has shifted from the textile to the car industry in recent years. The market is now consumer-oriented driven by individual demand, rather than producer-oriented determined by mass-production policy. Now the consumer demands individuality in textiles as well as in cars, reflecting the trend towards a higher level of economic life. For example, younger people are attracted by high-performance cars with particular specifications, for example, equipped with 4WD (four-wheel drive), a turbo charger or DOHC (double overhead camshaft). As a result the task of designing future cars will need to take regard of the individual and distinctive tastes of the consumer.

In cars, fiber materials are mainly used for the interior and tyre cord. Fabrics for the interior must have good physical properties such as weather resistance and easy processing, brilliant and deep colours, with lustre and a soft feel. Tyre cords are made of a composite material that should guarantee safety as well as driving comfort.

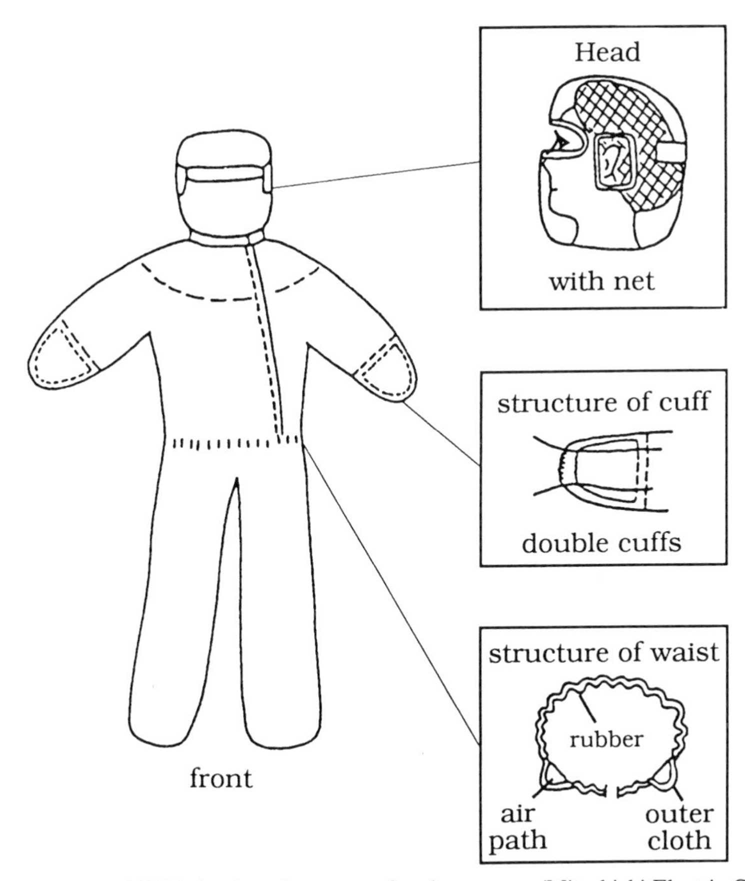

6.7 Fabrication of garments for clean rooms (Mitsubishi Electric Corp.).

The covering material for car seats is traditionally tricot, but moquette and other fabrics are constantly reducing the market share of tricot. Moquette now has a 30% share, which is almost equivalent to tricot (33%). Polyester is the main covering fiber (66%), followed by nylon (26%), which dominated the market a few years ago. Although small in amount, wool and acrylic fibers are also used.

Strong non-stretchable materials are preferred for the covering of car seats. Consequently, these are Jacquard-woven or dobby-woven. Elastic tricot fits the shape of car seats tightly without becoming loose, and many varieties varying from thick to thin-raised fabrics are available. Moquette fabric too has good stretch-resistance, and excellent dimension stability. More pile-woven moquette fabrics are employed for car seats because they provide a thick and heavy impression with a large pattern expression. Double-raschel or sinker pile-knitted fabrics are not so popular. However, thick but soft double-raschel is stretch-resistant and has a good dimension stability to fit the car seat effectively. This fabric also has a bright appeal with its small patterns. Sinker pile-knitted fabric is elastic, and gives a rugged impression with its large patterns.

Car seat fabrics undergo various finishing processes to improve their handling, touch, cleanliness and safety. Seats are often soiled while eating and drinking ingredients such as hamburgers, milk, coffee, juice, ice cream, etc. in the car. Soil-release finishing is thus undertaken using water- and oil-repellent treatments. Static electricity caused by friction between the passengers' clothes and car seats can place a charge on the human body, which can give an unpleasant shock when the traveller touches the car door. Some 60% of drivers have experienced such an electrostatic discharge, particularly when the air is dry. This electric shock can have an unexpectedly high voltage, over 3000 V. To some extent soil-release finishing suppresses static electricity on a car seat. However, antielectrostatic protection is normally achieved by coating the car seat with hydrophilic polymers. This can be achieved using blended fabric made from polyester and metal fiber and/or by mixing carbon or surfactants in the car seat packing material. Such antistatic treatment, however, is expensive and adversely affects the handling and light-fastness of the fabric.

Although each manufacturer competes in the design of car seats, seats must all be weather- and light-resistant for environmental reasons, and must be flame-resistant for safety. Cars are exported from Japan all over the world, where the environmental conditions can vary considerably. For example, high temperatures and strong ultraviolet radiation in a desert region might deteriorate the polyester or nylon fabrics of car seats, and cause the colour to change. This is known as photo-tendering. The covering material within a car must, therefore, conform to the highest degree of light and wet fastness. The flame-resistant treatment must meet the safety standard requirements. However, as the flame-resistance improves, the softness and touch is lost by adding more flame-retarding agents. Therefore, an optimum balance between design (including the touch and colour) and durability/safety must be aimed for.

6.5 Fibers in space

6.5.1 Structures in space

More than 30 years have passed since the first man set foot in space. Japan is now involved in the construction of a space station. The vacuum and weightlessness in space could provide assistance in producing new materials, whether crystals, metals or proteins. A small satellite will be used for the first series of investigations, but eventually a large space station will be able to house longer term experiments. Now fiber structural materials, such as CFRPs are being extensively applied in the construction of the space station. Apart from aluminium, which is used for sections where high pressure is exerted, CFRP composites make up most of the primary structures, because light weight is an essential requirement.

Japanese aeroplane industries, such as Mitsubishi, Kawasaki and Fuji Heavy Industries and certain electric companies such as Toshiba and Mitsubishi Electric Co. are able to produce such space structures to some extent. However, US industries are technically superior to their Japanese counterparts in this field. Dr Y. Fujimori of the National Aerospace Laboratory has recently pointed out that structural materials for a satellite and space ship must be lighter and more heat-resistant than the present CFRP composites. Atomic oxygen could burn and damage these materials during a low orbital flight at a height of 500 km. A US government report has shown that oxygen exists in its atomic form at this height, and could damage the surface of a space shuttle. Floating materials in space could also offer additional problems. A micrometeorite drifts at an average speed of 20 km/s. Bolts, nuts and fragments left by previous satellites drift at an average speed of 7 km/s. All these factors must, therefore, be taken into account when finally selecting the materials that make up the space station. The improvement of composites to meet these criteria is an active field in Japan at present.

6.5.2 Space suit

The Earth is surrounded with a thick layer of air that absorbs ultraviolet and other radiation; gravity is 1 *g* at the Earth's surface. Beyond this in space, any vehicle would be directly exposed to those radiations, but no gravity is exerted. The temperature in space is close to 0 °K (= –273.15 °C), whereas a spaceship's surface temperature exposed to the solar radiation can exceed 100 °C. A space suit, therefore, has vital functions in protecting the astronaut and assisting him or her in any activity necessary in the severe environment of space. It must, therefore, maintain temperature, humidity and atmospheric pressure by supplying oxygen and water vapour, and also protect against radiations from the galactic system, the van Allen belt, solar wind and solar flares, in addition to any hazard from micrometeorites, cosmic dust and atomic oxygen. For extra-vehicular activity, the requirements imposed on materials used are most severe. They must provide solar energy absorbency, resistance to heat and cold, gas permeability, thermal conductivity,

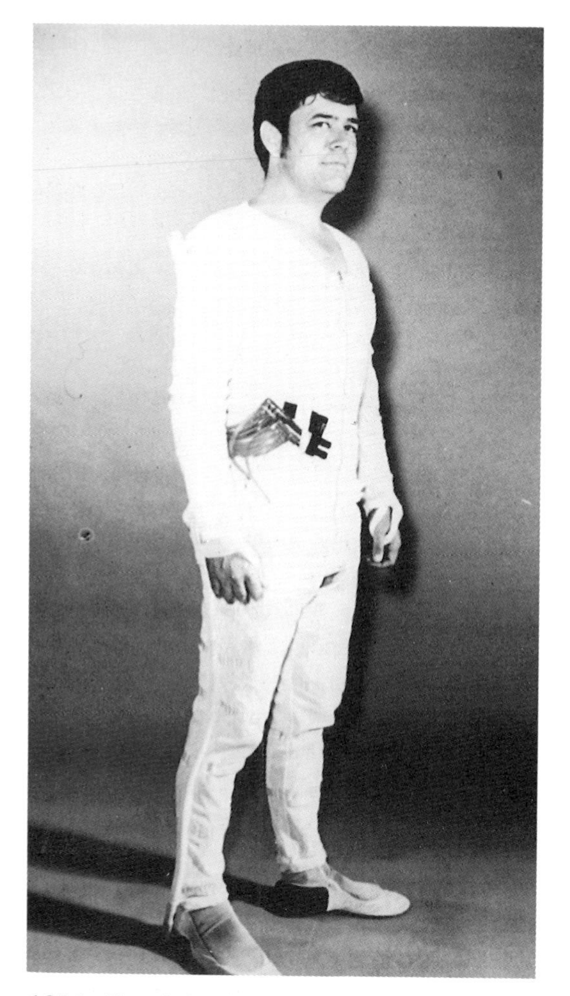

6.8 Liquid cooled underwear (NASA).

mechanical strength (both shock resistance and tensile strength), flexibility and processibility.

A space suit is essentially an extravehicular mobility unit comprising three sub-systems: the clothing section; the life-support system; and the communication/control module. The clothing consists of a suit, helmet and gloves/shoes. The suit and gloves are made mostly from various fibers. The suit includes water-cooled underwear, a second airtight garment and an outer protective coat. The helmet has included an extra-vehicular visor and an airtight helmet. The life-support sub-system includes the primary life-support system, a secondary oxygen pack, a drinking water tank and a urine collecting device. The water-circulated underwear controls body heat, the airtight

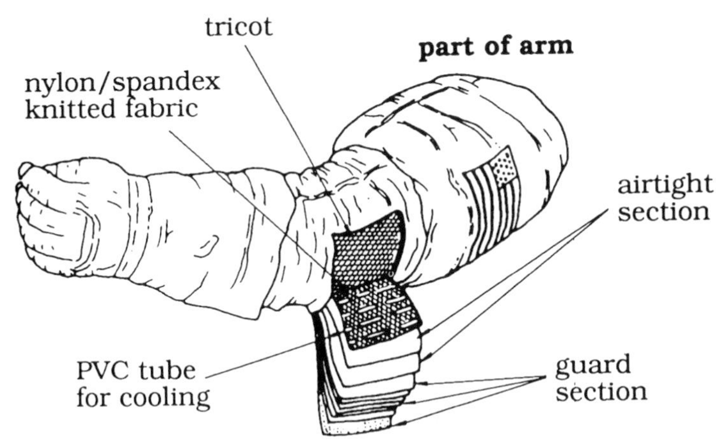

6.9 Structure of the arm of space suit (NASA).

garment maintains air pressure during work outside the spaceship, and a special cloth protects against radiations and floating micrometeorites.

The water-circulated underwear (see Fig. 6.8) has a three-layered structure made of tricot where it touches the body, 100 m (300 ft) of thin PVC tubes built into the middle layer and an outer covering of elastic nylon Spandex. Water runs through the PVC tube to neutralise the 500 kcal/h heat generated by the body while working. The airtight garment in turn is made up of three layers of nylon sheet net interlaced with neoprene-coated (synthetic rubber) nylon nets to prevent a blowing out of the cloth because of the inner pressure generated. The pressure within the airtight garment is maintained at 200 to 300 mm of mercury, so that no drastic pressure change is caused by body movements.

The radiation protecting cloth is made up of at least five layers to undertake various functions. The innermost layer is made of the nine-layer sheet of alternative aluminium foil and Dacron film to protect against radiation and insulate against heat. The next section is made of two layers of Kapton film of heat-resistant polyimide resin and glass fiber to ensure heat insulation, followed by the layer of Teflon-coated glass fiber sheet to insulate against heat and protect against micrometeorites. There is a final outer layer of Teflon (see Fig. 6.9).

The suit itself is divided into an upper and lower section. The former covers the neck, upper body and upper arms, and is coupled with the built-in life-support sub-system and control module. The lower section has a one-body structure from the waist to the shoes. The two sections are joined together by a connecting ring to maintain airtightness.

The Soviet space suit is a single compartment structure, and one enters the suit through the back door, which opens like a refrigerator. The complete extra-vehicular mobility unit weighs about 113 kg, which is too heavy for humans to wear on the Earth. In weightless space it floats like cotton. The cost of such a space suit is about US$650,000 with its development costing some US$25,000,000. Now the space suit comes in three sizes only, whereas in the Apollo programme suits were made individually for each astronaut.

6.6 Fibers and nuclear power

A number of organic materials are used in the field of nuclear power. Fibers are used extensively, not only as the basis for working suits, but also in scientific processing. For example, hollow fibers are an integral part of the filters used to purify the reactor's cooling water (Table 6.7). Fibers in composites can also be applied as structural materials. These have an advantage over metals since they do not themselves become radioactive when exposed to a neutron flux. The problem, however, is that in the radiation environment they suffer in their durability and mechanical strength. Composite materials are now being examined, such as glass fiber or CFRPs. For normal purposes they possess sufficient mechanical strength and durability, but their behaviour in an ionising radiation flux over a long period is not known. The maximum durability required for nuclear power equipment is about 40 years. Nuclear power stations are regularly inspected every year, so that all parts must have a guaranteed performance of at least one year, and their lifetime extended at least six to seven years. Thus, hollow fibers of vinylon and polyethylene, used in filters for cooling water are now being tested for radiation stability. The result depends on the operating conditions, particularly temperature, which is much more damaging when combined with ionising radiations.

Fibers too can be chemically and physically modified by radiation, a field that was intensively studied during the 1950s and 1960s. A new technique was recently developed to produce ceramic fiber by sintering organic fibers which had been treated with radiation. When untreated organic fibers are heated, they melt to give an amorphous state.

Radiation introduces cross-links between the polymer chains, without oxidation, and sintering thereafter gives ceramic fibers. For example, oxygen remains when polycarbosilane is conventionally oxidised and sintered to produce silicon carbide fiber, which decomposes below 1,300 °C. After cross-linking by radiation no oxygen is present, and the heat-stability of the resulting silicon carbide fiber improves by 200 °C to more than 1,400 °C. A similar technique can be applied to produce silicon nitride fiber by sintering the irradiating polycarbosilane fiber in a stream of ammonia. The silicon

Table 6.7. Use of fiber and radiation resistance required in a nuclear facility

Facility	Application	Radiation resistance	Remarks
Light-water reactor	Packing, valve, filter	2–0.1 MGy	In air
Fast breeder reactor	Insulator	300 MGy	In nitrogen
Nuclear fusion reactor	Insulator	50 MGy	At very low temperature
Nuclear fuel reprocessing	Filter, insulator, working wear	~0.1 MGy –10 MGy ~0.1 MGy	In air
Application of radiation	Vessel for irradiation	~10 MGy	In air

carbide or silicon nitride fibers can then be blended with aluminium or iron to produce a heat-resistant metal-like product. Since silicon carbide is semi-conductive and silicon nitride non-conductive, the resulting composite metal has quite distinct characteristics. In this way it should theoretically be possible to produce silicon carbide stable to 2,000 °C. If the heat-resistance can be improved to 1,600–1,700 °C, a new silicon carbide composite with ceramics can be produced.

The radiation-processing technique can also be applied to produce carbon fibers from PAN or pitch. Conventionally PAN or pitch fibers are heat-treated to oxidise them into non-melting cross-linked fibers, prior to sintering to produce carbon fibers. As a result, the pre-sintering product contains 10–15% oxygen, which converts to carbon dioxide as the polymer is degraded, so reducing the carbon fiber yield. Moreover, heat oxidation is a sophisticated process which is not easy to control on an industrial scale. However, the radiation modification can be simply performed at room temperature and requires no sophisticated control. The optimum radiation dose can be precisely selected in advance.

Two types of radiation sources are available at present: γ-radiation from ^{60}Co or fast electron beams from electron accelerators. Electron accelerators from Nisshin Electric Co. are now widely used in the electric wire and cable industry to cross-link the coating materials of thin wires in television circuits to prevent them melting when heated. The irradiation treatment can also be applied in paint curing to improve its adhesion and durability. The film industry uses this treatment also to ensure adherence of magnetic substances firmly on to videotapes.

The existing reactors are mostly constructed of metal (stainless steel), which give rise to radiation leaks due to corrosion or production of radioactive decomposition products. If fiber-reinforced composite materials can replace such metals, many of the problems caused by corrosion could be solved. Reprocessing facilities of nuclear fuels often suffer from pin-holes because of the chemical corrosion of the metal. This is another area where the use of composites is being explored. Finally, to store radioactive wastes, they are converted into a glass and stored. Composites could have a significant role here also. Indeed, heat-resistant organic composites could, in the future, replace metal in many parts of nuclear reactors and in the associated shielding.

6.7 Fibers in sport

Sports records have improved and so has the associated sportswear. Modern athletes require that it should be functional and fashionable. Indeed, sportswear has contributed not insignificantly to the steady improvement in athletic records. Highly functional fiber materials, including composites, have extensively been applied in sportswear, in the construction of athletics stadiums and to assist athletes achieve a better performance.

6.7.1 Highly functional materials in sportswear

Sportswear needs to be pleasant to wear from a physical and functional standpoint. It must not restrict physical movement, or prevent sweating and should be fashionable. Sportswear also finds use as leisure clothing, which considerably extends its scope. Here we consider recent trends in sportswear materials.

6.7.1.1 Sweat- and moisture-absorbing materials

To be comfortable, sportswear needs to be able to absorb moisture and sweat. Teijin has developed a porous polyester fiber Wellkey, which has a hollow centre, with a large number of micropores at the surface. The micropores of 0.01–0.03 mm in diameter are homogeneously distributed, and some run right through into the hollow part (see Fig. 6.10). Sweat is immediately absorbed through pores and diffuses into the hollow centre, so keeping the fiber surface dry. Sportswear made of Wellkey, causes no chilly sensation on sweating, nor does it stick to the skin and restrict body movement. The Wellkey uniform, designed by the Assics Co., has been adopted by the All Japan Women's Volleyball Team.

6.7.1.2 Moisture-permeable water proof materials

Gore-Tex is moisture-permeable and waterproof and first appeared on the Japanese market in 1977. It is made from a microporous polytetrafluoroethylene film laminated with cloth. Since then, many materials with similar functions have been developed and used for outdoor sports, such as golf or ski-wear. Special mention must be made to Exceltech, developed by Unitika,

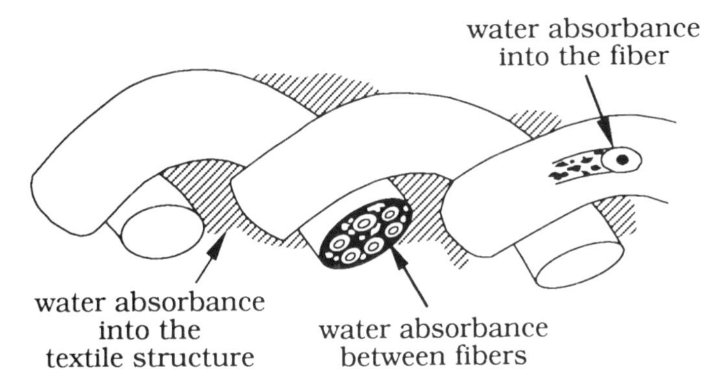

Water absorbance mechanisms

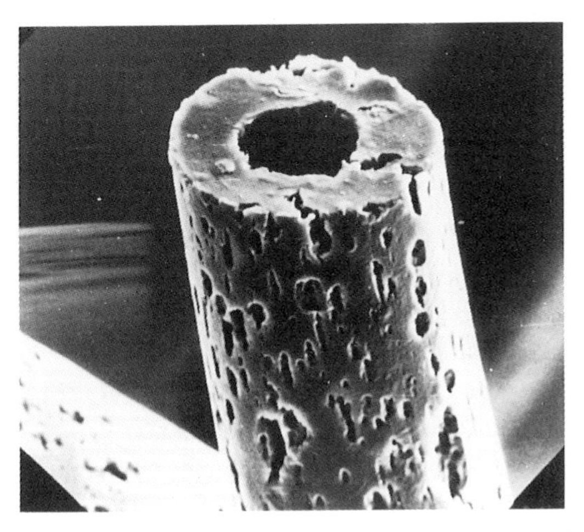

6.10 Water absorbance of the porous polyester fiber Wellkey (Teijin Ltd).

which is finished by wet-coating with polyaminoacid to utilise the high moisture permeability of this polymer. Water-repellent Super-Microft (developed by Teijin) also comes into this category of functional materials, as described in Chapter 4.

6.7.1.3 Low-air-resistant clothes

Speed or flying distance is of primary importance in downhill and ski-jumping competitions. A fabric of low air resistance, Cosmopion, was developed by Toray, in collaboration with the Laboratory of Aeronautic Technology, Kyoto University, in response to demands from these fields. Cosmopion is a two-way tricot, with an extremely flat surface of densely knitted ultra-fine nylon fibers. When used for ski jumpsuits, the flying distance was found to be much greater with this flat surface (like an aeroplane) compared with a dimpled surface (like a golf ball). In Cosmopion the height of the needle loop is minimised and the space between loops reduced by expansion in a lateral direction, as a result of the low-modulus characteristics of ultra-fine nylon fiber of 8 mm in diameter. In this way an extremely flat surface is obtained. Cosmopion has proved its worth at the Olympics and in world cup ski-jump competitions, and will surely now be used for other sportswear which require speed, as in downhill skiing and bicycle racing.

6.7.1.4 Other functions

Specific characteristics can be imparted into fabrics to enable them to be applied for specific functions. The improvement of swimming records is partly due to the development of a thin and elastic swimming suit made of a tricot fabric comprising nylon and polyurethene, which functions as if it were a second skin. The swimsuit used for water polo is coated to make it slippery, thus preventing the player being caught during the match. Dyes and other materials used for the swimsuit must also be chemically resistant to the combined effect of chlorine used for the sterilisation of pool water and light.

A fabric produced from kapok seeds is employed for the fabrication of life-jackets. It is light, and is only a third of the weight of cotton with the same volume, but its buoyant force is 30 times as large as its own weight.

A motorcycle racing suit is made of leather or artificial leather, made up of very fine fibers. The cyclist wears a fire-proof wool shirt under the suit and underwear of aramid fiber in case of an accident.

Other specific characteristics desirable for ski-wear are heat insulation, moisture permeability, waterproofing, windproofing, and attractive fashions. Air permeability particularly affects downhill speeds. A downhill suit used in the Alpen race must conform to an air permeability of 30 l/m^2 s as specified by ISF (International Ski Federation) regulations (50 l/m^2 s up to 1984).

Sailing boats are now built of lighter but mechanically stronger composite materials. Water-repellent polyester or high-tenacity polyethylene is used for the sail, which will dry very quickly after an encounter with a wave.

Fishing rods are mostly made of GFRP (glass-fiber reinforced plastic), with the fishing gut of nylon or polyester monofilament. These have completely replaced the bamboo pole, horsetail hair and silk thread used until some 20 years ago. Carbon fiber reinforced or ceramic whisker reinforced plastics are also being increasingly used for fishing rods. Synthetic fishing gut is processed in a manner that ensures that it is invisible and does not repel fish. This is a field that has been revolutionised by the use of new materials.

Most of the tennis gut used now is made of nylon, which has much better durability and easier care compared to the sheep or whale gut used previously.

Polyoxymethyl gut has also been developed recently using high-tenacity polyoxymethylene fiber (see Chapter 3).

A parachute for skydiving must be light and strong. Paragliding is now becoming popular, and the needs here are essentially the same as for parachutes from a functional viewpoint. Nevertheless, better gliding ability needs to be added to paragliding fabrics. These are made from nylon, as are the attached ropes and safety belts.

Nylon ropes are also mostly used for mountaineering, rockclimbing, ice-climbing and caving. Polyester ropes are also in use, depending on the location of use and purpose.

6.7.2 Fiber composites in sports goods

CFRPs have now found extensive application in sports goods, and in Japan this is the major area of their use. The first application of CFRP was for pole vaulting, which was first made of bamboo. After replacement by CFRP, world records have consistently improved. Sailing boat bodies and wind surfing boards were originally made of wood and later foamed resin, but now GFRPs or CFRPs are used to produce better performances. CFRP was employed for archery bows soon after it was first introduced, but recently the GFRP bow has appeared. Ski plates are made of laminated sheets of fiber reinforced plastics of glass, carbon, aramid, silicon carbide or alumina fibers.

Sports and leisure goods make up almost 70% of total CFRP products in Japan, and of these fishing rods, golf shafts and tennis rackets make up 90%. Composite materials are suitable because they are light, highly elastic, have excellent mechanical stability and are flexible from design considerations. These many required characteristics could not be achieved with conventional materials. For example, a CFRP golf shaft can now be designed to have the required degree of shaft twist, flexibility and balance. A CFRP golf head can be made with a wider sweet spot. A CFRP fishing rod can be made light and easy to handle by designing in the optimum balance. At present CFRP has a 20% share of the golf shaft and fishing rod market. Inevitably, in future more CFRP will be used for goods requiring high-performance characteristics.

6.8 Fibers for geotextiles

Since the Tokyo Dome Stadium first utilised synthetic materials for its ground, the artificial lawn is gradually replacing natural grass in football and baseball grounds in Japan. These artificial surfaces are made from nylon, PVC or polyethylene (PE) fibers planted on to polyester mesh. A sand-filled artificial lawn was developed for the tennis court which could provide comparable surface characteristics to that of a natural lawn, for example, the height of the bounce of a ball. The special mat used as a launch pad for the high jump or pole vault is filled with polyurethane foam. Such applications of fibers fall into the category of geotextiles, which includes fiber-processed materials used in construction and building. Fibers are now being increasingly used in the construction industry in Japan to counteract the shortage of natural materials. Although geotextile is a new term, made up of geo (the earth) and textile, the use of such textiles can be traced back to a 4,000-year-old arch in China constructed with a clay earth mixed with fibers, or the Great Wall built 2,000 years ago.

The textile materials used as geotextiles can vary from fibers (long filaments and short cut fibers) to fabrics (knitted fabrics, woven fabrics and non-woven fabrics). They function as a reinforcing material for soil or cement, or as a filter in drainage material to separate water and remove fine clay suspensions. Usually, however, additional strength is needed and fibers are particularly suitable since a fibrous structure is characterised by molecular orientation in the direction of stretching, resulting in producing high tenacity in one direction. Another application is for high-tenacity fiber net covering a steep embankment to prevent landslips.

The International Geotextile Society was established in 1982 to promote new applications of geotextiles, and in Japan the Geotextile Research Group started its activity in 1985 within the Society of Fiber Science and Technology. A questionnaire sent out by the Japan Chemical Fiber Association has shown that in the various categories, the demand for geotextiles has increased between two and five times in the past five years. The total world demand in 1985 was 340 million m^2 and exceeded 800 million m^2 in 1990.

Most geotextile materials are light, strong but cheap and are mainly fabricated from polyolefins such as PE and polypropylene (PP). Nylon, glass and polyester (PET) are also used, and the super-fibers (aramid, carbon and high-tenacity vinylon fibers) are gradually increasing their market share. Carbon and aramid fibers are extensively used in Japan for reinforcing concrete. Steel fibers, used previously, are heavy and less flexible whereas glass fiber is too brittle. Super-fibers, eliminate such weak points.

The outer curtain wall of the 37-storey Arc Hills Mori Building in Tokyo (Fig. 6.11) is constructed from 5,540 tonnes of CFRC. This reduced the wall weight by 60% and the earthquake load by 12% compared with conventional constructions. The total weight of steel frame supporting the building was accordingly reduced by 4,000 tonnes to 20,000 tonnes. CFRC is made of a mixture of short cut carbon fibers (3–10 mm in length and 15 μm in diameter) suspended in cement to a concentration of 2–4% volume. Compared with

6.11 CFRC curtain wall of the Ark Hills Mori Building in Akasaka.

ordinary concrete, it has better mechanical properties, durability and size stability. High-tenacity carbon fiber is heat-resistant and chemically stable in the strong alkali conditions of concrete. Since it results in such an overall weight reduction of a building, the use of the carbon fiber is commercially viable even though pitch-type carbon fiber is expensive and currently costs about US$33 per kg.

The Kajima Corporation first employed CFRC as the material for curtain walls. The company is now investigating the practical use of high-strength composite concrete structural materials using aramid fibers developed by the Teijin Co. Here the aramid fibers are knitted into a bundle and fixed with resin to produce a rod-shaped material. This possesses a higher mechanical strength than an iron rod and can, therefore, replace an iron bar in ferro-concrete.

The Mitsui Construction Co. has developed an aramid fiber reinforced concrete, in cooperation with Toray. Here the mixture of cement, sand, water, mixing agent and short cut fibers of 5–15 mm in length is extruded, after which a number of rod-like knitted aramid filaments are inserted into the cement mixture.

Kuraray has established the technology to produce vinylon reinforced synthetic slate as a substitute for asbestos. It is now supplying this product to the world's largest asbestos-cement manufacturer, Eternit Co. Ltd. Unitika Chemical Co. of Japan has a joint venture with Turner and Newall Co. UK, to produce the vinylon synthetic slates used for roofing in Europe.

Although membrane-type materials have been used as air dome structures for large spaces over temporary exhibition pavilions for short-term use, the durability and heat resistance of these materials was not considered appropriate for permanent buildings. Glass-fiber fabric coated with fluororesin was developed for this purpose. It is suitable as a permanent roofing material because of its durability, heat resistance and mechanical strength. The first use for air dome permanent roofing was in the Reiyukai Mirokusan built in 1984, followed by the Tokyo Dome Stadium, completed in 1988 (see Fig. 6.12).

In 1985, the total textile consumption in construction was 9,237 tonnes woven/knitted fabrics and 9,080 tonnes non-woven fabrics. The volume of non-woven fabrics had increased by 50% since 1980 whereas the quantity of woven/knitted fabrics had remained unchanged. This growth has continued into the mid-1990s. Non-woven fabrics have proved better for filtration and separation, and woven/knitted fabrics are better for reinforcing purposes. The following examples will demonstrate the present variety of geotextile applications.

Enka Co., The Netherlands, developed a railway sleeper bed using high-modulus and good ultraviolet-resistant polyester fabrics. These stabilise the road bed and prevent mud scattering. Enka claims that the use of this sleeper bed reduces the long-term deformation (plastic strain) of rails. Sponge-like non-woven fabrics of nylon filaments have been experimentally paved under the railway line from Tokyo to Shinagawa in the Tokyo Underground to absorb the vibration.

Woven/knitted fabrics are commonly employed for draining. A polypropylene- or polyester-fabric bag filled with sand is used to drain water and so reclaim soft wet ground. It is known as the packdrain method. The French National Research Center for Public Works has recently proposed the Texsol engineering system to reinforce soft ground by jetting and entangling polyester filaments into the ground using high pressure water. Hoechst AG, Germany, has developed a PET-fabric wall, which costs only 30–40% of that of a conventional concrete wall. A cylindrical or pillow-shaped knitted bag

6.12 Tokyo Dome alias "Big Egg", in the Korakuen Stadium.

has been developed for bank protection by Karl Mayer AG, Germany. ICI, UK, supplies net-type PET woven or knitted fabric sheets to cover sand banks and prevent wind erosion in dry areas. The Asahi Chemical Industry Co. Ltd has developed a concrete moulding which allows the removal of water from the filler using a nylon fabric bag possessing good water permeability. The Kumagai Gumi Co. applied layered water-permeable fabric sheets to plastic liners to mould concrete which then has good surface strength without pitting. Du Pont, USA, markets the polyethylene non-woven fabric Tyvek to wrap around houses to provide thermal insulation and prevent dew-condensation.

These are only a few examples to demonstrate the rapidly expanding application of the textile materials in construction. It can be anticipated that inorganic and metal fibers will soon be incorporated into geotextile materials to satisfy the increasing technical requirements. If engineers and designers appreciate their potential the annual demand for textile materials could increase to 20 million m^2 within a few years.

6.9 Fibers in the ocean

The days when the fishing net represented the main use of fibers at sea have long since disappeared. They are now widely used on the sea, in the sea, and below the sea. A large sailing vessel remains a symbol of past glories of the sea. However, such boats have now moved into the modern era. The first energy-conserving tanker with two laminar-flow type hard sails controlled by electronic systems was launched in 1980. Now 15 energy-conserving sailing vessels are in service (see Fig. 6.13). The application of modern electronic technology has made possible the revival of large sailing vessels on a commercial basis.

Small sailing boats are also still popular for sporting and fishing. Most of these boats are now constructed from fiber-reinforced plastics (FRPs) which last longer than steel and wood. Indeed, the biggest problem is their disposal when their usefulness is over. Today PET plain fabric, coated with PVC, is a main material used for sailcloth.

FRP was extensively used in the construction of the deep-sea (6,000 m below the sea level) submarine *Shinkai* to provide it with the necessary buoyancy.

Some 30 years ago, a membrane-structured tanker was built in the UK. Now the Institute for Industrial Products, Japan, is investigating the possible application of large membrane-structured containers to store oil, drinking water or foods on or in the sea. There is a plan to stock rice at the bottom of the Biwa Lake, Japan, where the temperature is low (4 °C) and constant, and ideal for the storage of foods over long periods. These soft containers are constructed from a nylon or polyester membrane base and coated with synthetic rubber. When an oil spill occurs at sea, it is necessary to set up an oil-block to enclose and restrict the movement of the oil. Such oil-blocks are constructed by joining together flexible fences of glass fiber fabric coated with resin.

Many textile or membrane structures can be found even within the sea. Membrane screens have been built deep in the sea to produce an upward sea current. This directs the plankton to near the sea surface, and hence enhances the value of a fishing ground. An artificial seaweed bed has been constructed from membranes made of synthetic fiber fabrics. Such a bed has proved good fish breeding ground, providing a location where seaweeds and plankton can grow and where fish can live. As a result the marine ranch business is developing quickly through the application of fiber technology.

Pollution from moving mud can damage fishing grounds during dredging operations. To prevent this occurring, a membrane fence (silt curtain) is set up at the sea bottom near the dredging area to reduce the mud spreading by changing the current direction and reducing current speed. Textile barriers are also spread over the sea bottom to prevent sand from flowing out from reclamation works.

For such purposes the membrane structure has many advantages compared with other structures. It is light, thin, flexible and spatial, and can be folded

6.13 An energy-conserving sailing vessel.

into a compact size to store or transport. Production is relatively easy, and its shape and dimensions can be designed readily, according to requirements. Modern fiber technology has made it possible to construct a vast variety of products. There are wind-resistant (against a 60 m/s wind speed) membrane structures, very long structures, extending over several tens of kilometres, tube-like structures and the vast tank-type containers up to millions of cubic metres. Their flexibility is so extensive that they are now proving indispensable for ocean developments.

7 New high-tech fibers and *Shin-gosen*

Synthetic fibers were first produced as imitations of natural fibers. The desire to create a fiber with a silk-like touch was the trigger in the production of many new kinds of synthetic fibers. Today, the technique of imitating nature (biomimetics) has improved to such an extent that not only can the basic structures of living things be duplicated, but the more precise functions of living organisms can also be copied. Using the concept of biomimetics, synthetic fibers with specific properties have been produced, which in turn have enriched our lives.

Rayon was invented in 1884, nylon in 1938 and *Shin-gosen* with silk-scrooping sound emerged during the 1980s. All three types of fiber arose because of the desire of the researchers to create silk artificially. Despite the changes in society over the years, the appeal of natural silk has remained, which has led to the production of better silk-like artificial fibers.

The success of the products which have been based on the natural system is driving an even more intensive utilisation of biomimetics. Nature will continue to be challenged in the quest for new fibers to meet the evolving needs of society. High-tech fibers are, therefore, continually moving on. Although previously described, in this chapter the newer technologies and developments will be given.

7.1 Various categories of high-tech fibers

In the spring of 1985 a book entitled *High Technology Fibers* was published in the USA, resulting in the term "High-tech fiber" being adopted. A high-tech fiber can be defined as a fiber produced by high technology, namely, new fibers having superior properties to ordinary fibers, and can be split into three categories: high-performance fibers, high-function fibers and high-sense fibers.

7.1.1 High-performance fibers

Fibers need to possess mechanical properties such as tensile strength and modulus in some degree, as well as thermal resistance, good dyeing property, weather and chemical resistance. High-performance fibers by definition need to have improved properties compared with ordinary fibers. High-performance fibers with extremely high tensile strength and modulus are referred to as super-fibers. These can be used as ropes by themselves, but are more often used as reinforced fibers in ACMs in the field of aerospace industry, for tennis rackets, golf shafts. Typical examples of super-fibers are PAN-based carbon fibers, *para*-type aramid fibers, high-tenacity and high-modulus polyethylene fibers, and polyphenylene bisoxazole (PBO) fibers. The tensile strength of fibers is shown in Fig. 7.1. It illustrates how many kg/mm^2 cross-section of super-fibers can sustain, in comparison with ordinary fibers. More than a ten-fold increase in strength can now be achieved.

7.1.2 High-function fibers

Fibers that utilise special morphological characteristics to function are termed basic function fibers, or primary function fibers. They are designed to fulfil a specific functional need such as being moisture permeable or water repellent, highly absorbent to water, antibacterial and deodorant. When they are ultraviolet resistant or heat storing, they are sometimes referred to as secondary function fibers. Of course, these terms are not mutually exclusive.

The technologies that add high function to fibers are shown in Table 7.1. These are often used in combination, particularly technologies (1) and (4). *Shin-gosen* is a name given to fibers that combine all the technologies from (1) to (4), which are combined also with innovative sewing methods.

7.1.3 High-sense fibers (or aesthetic fibers)

High-sense fibers are fibers that are extremely comfortable to wear, are highly

Table 7.1. Technologies to introduce "high functions" into fibers

Technology to modify fibers	Functions introduced
1 Modification and improvement in materials	Antipiling, antistatic, hydrophilic, flame-retardant
2 Innovation in spinning	Hollow fiber, non-circular cross-section, composite fibers, ultra-fine fibers
3 Improvement of fabrics*	Bulkiness
4 Modification and improvement during processing	Anti-static, sweat-absorbent, waterpermeable/waterproof, flame-retardant, antibacterial/deodorant

* Textile, knit, and non-woven fabrics

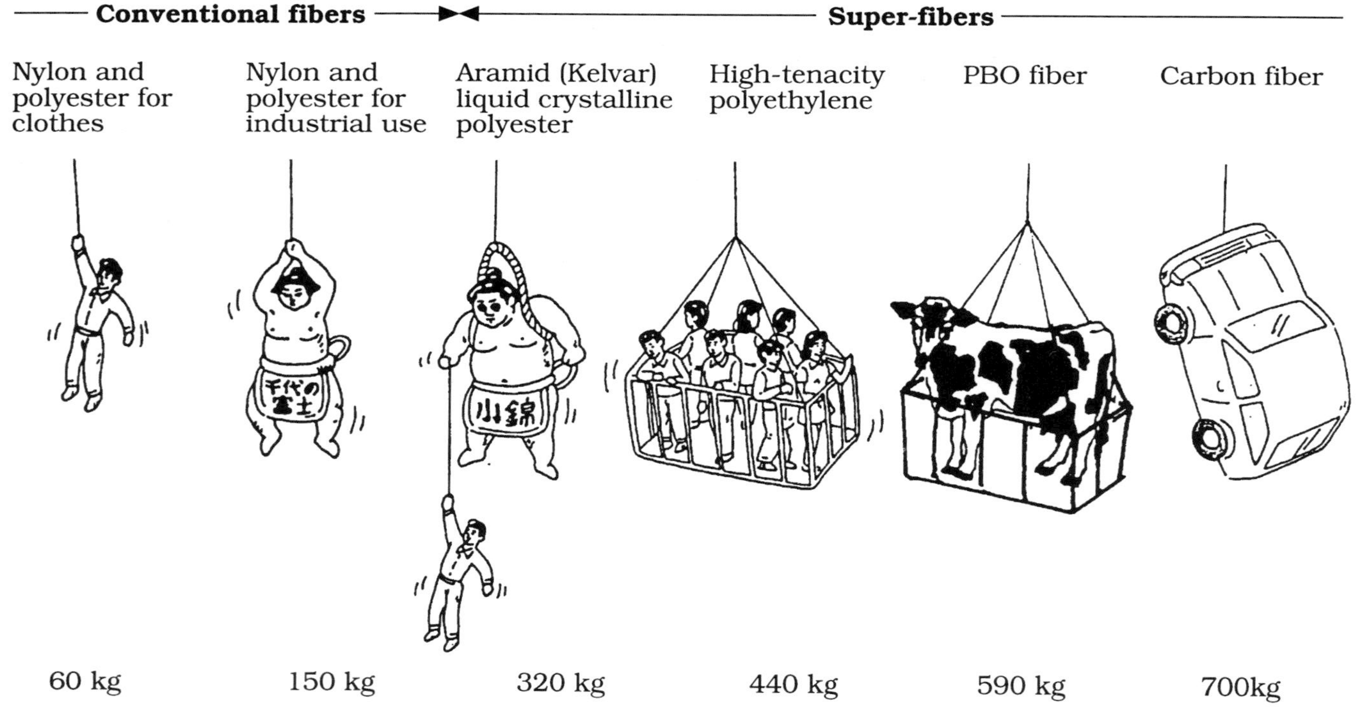

7.1 Super-fibers: how many kilograms can a fiber with 1 mm2 cross-section hold?

Table 7.2. Various aesthetic fibers produced by changing the cross-section

Trade mark	Producer	Cross-section	Basic technology	Specialities of the products
Solo Sowaie	Asahi Chemical Industry Ltd		Hollow, triangular, thick and thin	Higher bending stiffness, mild colour
Fontana μ	Asahi Chemical Industry Ltd		W-shaped, self-crimping	Bulky, crispy, dry and cool to touch
Soielise N	Kanebo Ltd		Pentagonal cross-section	Mild lustre, dry to touch, water-absorbent
Vivan	Kanebo Ltd		U-shaped cross-section, thick and thin	Mild lustre, dry, spun-yarn-like, higher bending stiffness
Dephorl	Kuraray Ltd		Flat cross-section, self-crimping	Deep colour, bulky, higher higher bending stiffness
MSC	Unitika Ltd		Arrow-like cross-section	Dry and cool to touch
Mixy	Unitika Ltd		Random and multishaped cross-section	Dry to touch, natural appearance, higher bending stiffness

fashionable, with superior aesthetic and sensual factors due to their hand feel properties. Owing to their ultra-fine character and non-circular cross-section, these fibers provide fineness, fullness and softness, lustre, crispness and drapability comparable to silk. *Shin-gosen* is one of the high-sense fibers. "Sense" here is used to indicate that the new *Shin-gosen* has added an aesthetic quality, not previously achieved. Examples of how are shown in Table 7.2. Changing from the conventional circular cross-section can change the overall properties of the fiber. Non-circular cross-sections give the fiber, not only a different lustre, but also a remarkable change in bending stiffness and handle. A regular triangular cross-section gives a fiber with 1.2 times larger bending stiffness.

7.2 Development of *Shin-gosen*

The word "*Shin-gosen*" was introduced by the media in Japan during the latter half of 1988 to describe a completely new generation of textiles based on synthetic fibers. Thereafter, there was a "*Shin-gosen* boom" and the term was adopted elsewhere, without translation or clear definition. Such fibers must also be regarded as High-tech, produced by high technology. Micro, random and a combination of all available technologies were used to produce these fibers, which have a different quality and performance from those of ordinary fibers.

7.2.1 Introduction of Shin-gosen

The reasons for the "*Shin-gosen* boom" are as follows:

1 Consumers were demanding change in fashion and the changes that had started in the early 1980s had reached the limit of the development of the materials.
2 Every synthetic fiber manufacturing company tried to develop highly value-added new materials, such as high-sense materials, with unique quality and taste and produced new materials, particularly for women's clothes.
3 The development and combination of technologies ranging from fiber production to processing of silky polyester produced *Shin-gosen.*

Peach-skin-like materials became a trigger for the boom because of their new appearance and hand touch. Next, fabrics made of strongly twisted composite fibers were used for suits and jackets, which had usually been made from wool. Finally, high quality, fine texture and bulkiness of new silky materials made of polyester became popular for blouses and dresses. The success of *Shin-gosen* came from the enlargement of this market, not only in blouses and dresses, but also for jackets and suits. Common features of three types of *Shin-gosen* are peach-skin quality, new silkiness and a worsted character giving fullness and drapability, which provided a high quality awareness and gained consumers' satisfaction.

7.2.2 Technological features of Shin-gosen

Shin-gosen is produced by combining several established and new technologies ranging from polymer production to sewing (See Table 7.3). Three areas of progress can be identified. The first is the production technology for specific polymers, which could be new co-polyesters, and as such were designated "new polyesters" to distinguish them from the ordinary polyesters. The second covers spinning, weaving and dyeing technologies, to produce non-circular cross-section fibers, ultra-fine fibers, mixed spinning and surface treatment. The third one is new sewing technology to produce completely new fabrics. Therefore, *Shin-gosen* may be defined as fabrics for clothes that have completely different taste and functions compared with those prepared from conventional synthetic and natural fibers. Additionally, the development of *Shin-gosen* required not only the application of a set of new combined textile technologies, but also another technology to evaluate overall sensitivity of need, combined with the engineering skills to achieve a fabric with its own in-built aesthetic sense.

Table 7.3. Established technologies and new technologies to produce *Shin-gosen* fabrics

Process	Previous technologies for first and second generations	Newly developed technologies for the third generation: the "*Shin-gosen*"
General	Single technology or its simple combination	Combination of the plural technologies to create highly refined products
Polymerisation	Bright (non-pigment) Dull (pigment) Cationic dyeable polymers (sulphonate group)	High-shrinkable copolymers, high-gravity polymers, unevenly degradable polymers Easily-degradable polymers
Spinning and drawing	Trianguar cross-sections	Specialised, uneven and randomised cross-sections
	Thin fibers	Self-extensible fibers, super-low shrinkable fibers, high-shrinkable fibers, specialised conjugated fibers, super-fine fibers
Blending filaments	Simple combination of heat treated and untreated filaments	Combination of super-low shrinkable filaments, super-less-shrinkable filaments and self-extensible filaments
Texturing	Specialised false-twisting	Composite (mixed) false twisting sheath-core or double-layer false twisting
Dyeing and finishing	Alkaline reduction	Splitting conjugated fibers Heat treatment of high-shrinkable fibers, less-shrinkable fibers and self-extensible fibers, mild pile raising, surface treatment of fiber, randomising

7.2.3 Development of technologies to produce silk-like fibers

Synthetic conventional melt-spun fibers produced in the 1950s had a circular cross-section, which gave the fabrics a flat hand touch like paper. To improve the hand feel, technological developments led to silk-like, cotton-like and wool-like fabrics. Figure 7.2 outlines the stages in these developments, from the birth of synthetic fibers to silk-like fibers and eventually *Shin-gosen*.

The first generation started around 1964 by reproducing the silk triangular cross-section, which imparts a silk-lustre and silk-crispness to the fibers. During the production of natural silk fibers, sericin is removed from the raw silk thread by alkali treatment, leaving only the fibroin. To mimic this process, the polyester fabrics were reduced in weight (*ca.* 25%) by caustic alkali treatment. As a result the pressure between fibers within fabrics decreased, so that silk softness, and drape were introduced into the fabrics. The steps as in Fig. 7.3 can be represented as:

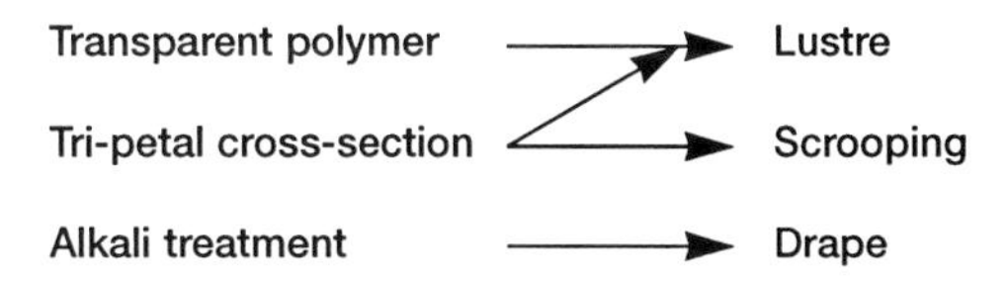

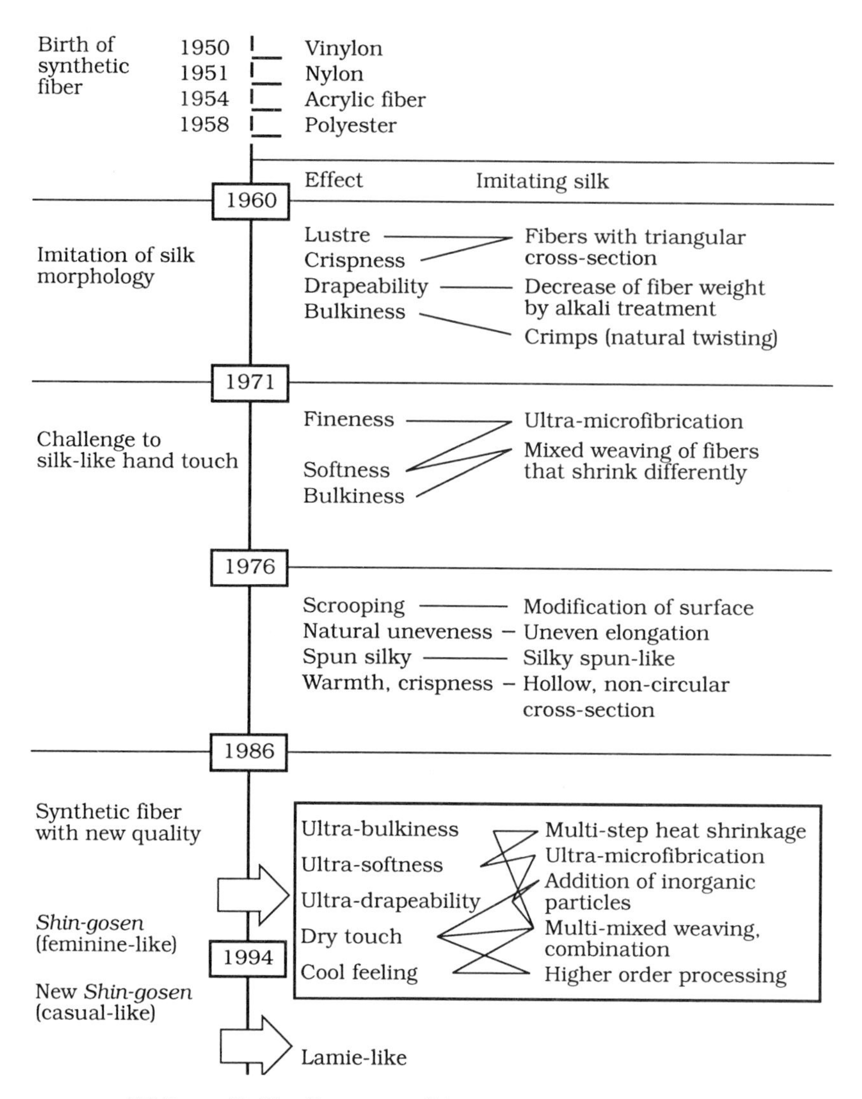

7.2 From silk-like fiber to new *Shin-gosen*.

The second generation, to reproduce silk hand-touch started around 1975. Here new technology gave bulkiness and softness to the fiber by the development of mixed spinning with different types of fibers having different shrinkage levels.

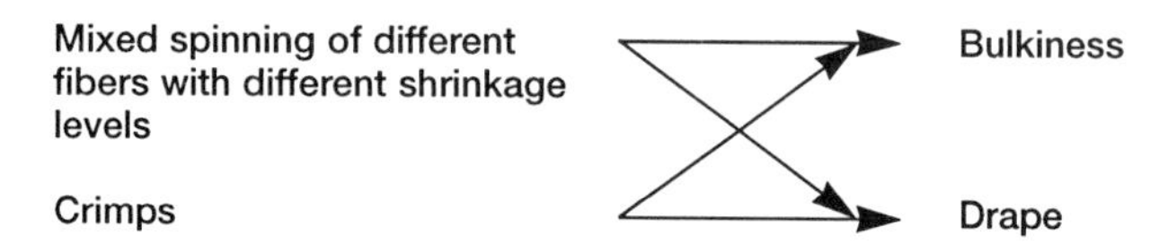

The third generation to seek natural taste and unevenness started around 1979. Demands for variability and higher quality stimulated research to produce materials with a more natural appearance and touch. As a result, fibers were produced with complex unevenness, such as multi-grooves with thick-and-thin, and uneven cross-sections, which synthetic fibers did not have because of their uniformity (Fig. 7.4):

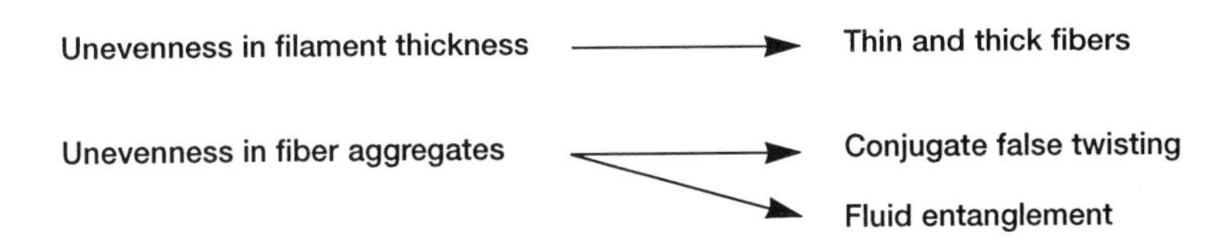

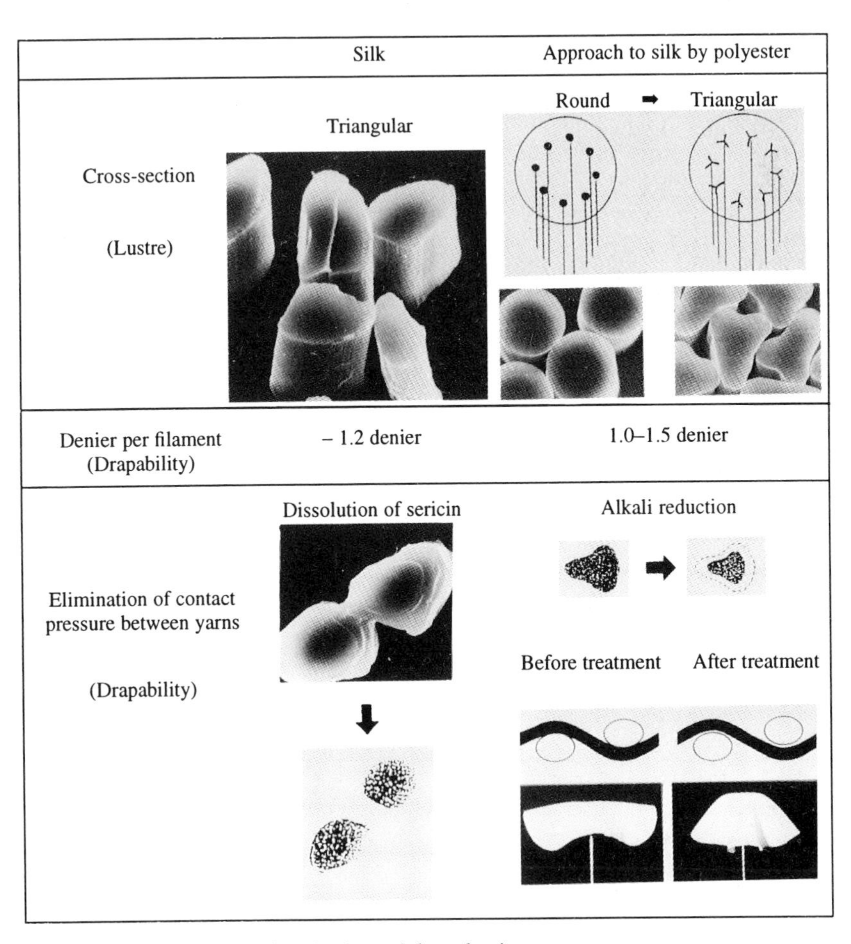

7.3 Triangular fine denier weight reduction.

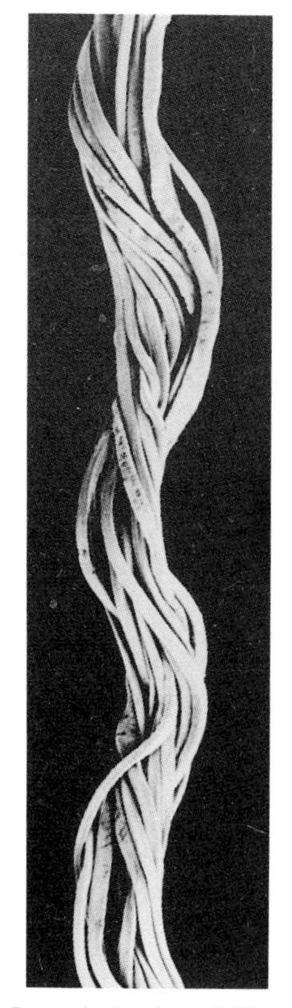

Thick and thin fibers

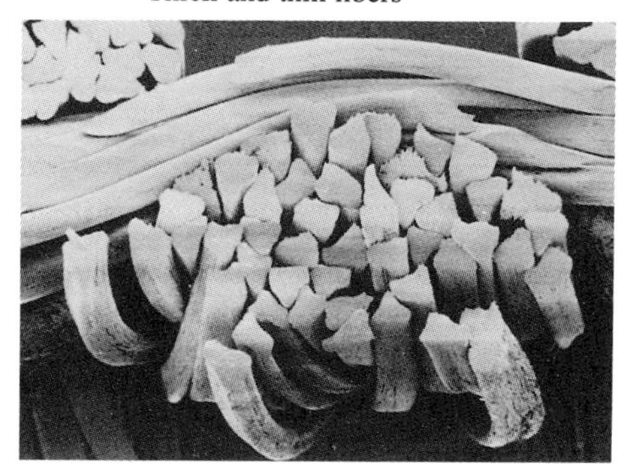

Irregularly shaped fibers

Fabrics of irregularly shaped fibers

7.4 Irregular shape fibers.

Motivation of the fourth generation was to produce synthetic fibers with a unique touch, dyeability and function, and this activity started around 1986. Within a short time, manufacturing of *Shin-gosen* started on a full scale. These were synthetic fibers whose quality and touch could not be provided by normal synthetic fibers and natural fibers, and which introduced a new dimension to fiber design.

7.2.4 Shift of paradigm in the development of Shin-gosen

Looking back, it appears that concepts in fiber development have changed about every 11 years, which has been the result of long-term research, a limit reached in the development introduction of new technology to produce new materials and the changing needs of society. These changes are outlined in Fig. 7.5. The various milestones are worthy of note here.

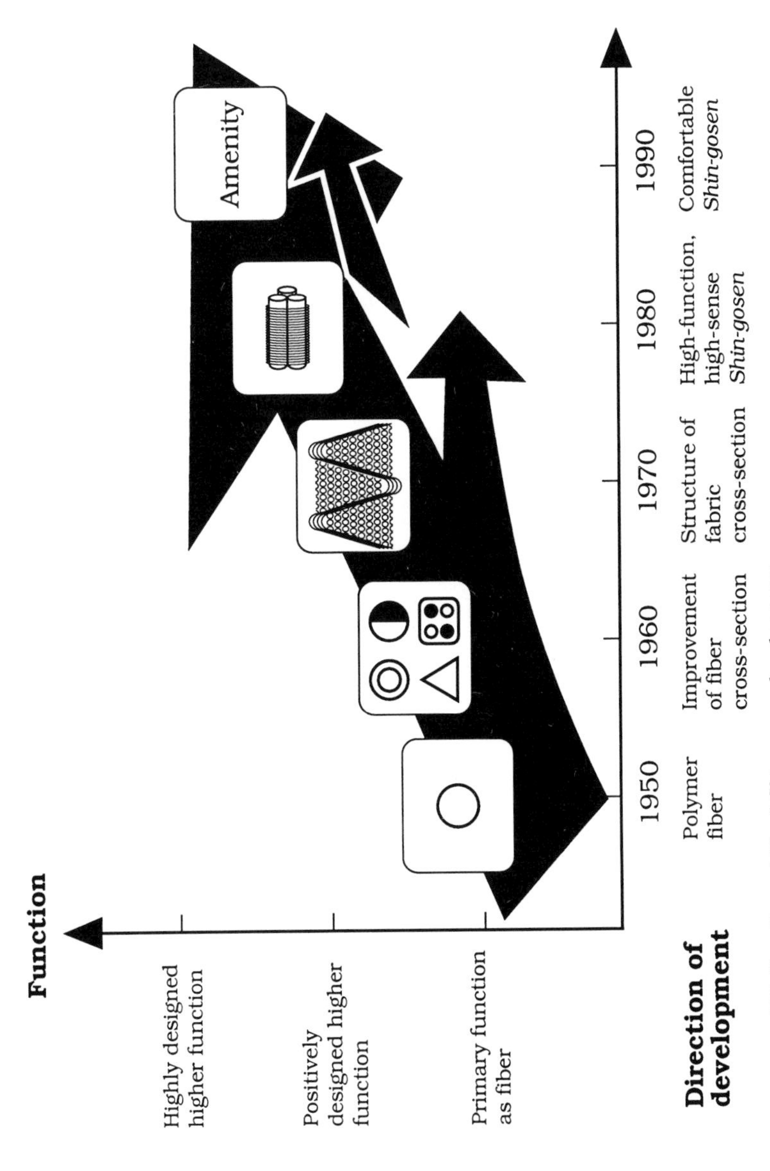

7.5 Paradigm shift of *Shin-gosen* development.

7.2.4.1 Production of fibrified polymer during the 1950s

Production of fibrified polymer started in Japan in 1950, 1951, 1954 and 1958 for vinylon, nylon, acrylic fibers and polyester, respectively.

7.2.4.2 Improvement of cross-section of fibers during the 1960s

The fibers with triangular cross-section were developed to imitate the cross-section of silk. Research on bilateral structure of wool fibers led to the production of composite fibers by composite spinning technology (bound type and core–sheath type), and fibers based on the structure of cotton and hollow fibers were devised. Following these developments, important new fiber materials were developed.

7.2.4.3 Improvement of cross-section of fabrics during the 1970s

The hand touch of fabrics is controlled not only by the cross-section of fibers, but also by the distance between fibers in fabrics. Artificial leather with a suede touch was developed in the 1970s. The technologies to produce bimetal-type composite fibers and core–sheath composite fibers are used for the production of such fine microfibers. A sheet of microfibers can be produced by "sea–island" technology, when in a multi-component system, one of the components ("sea component") is dissolved away. When the fabric has been formed, the surface of another component can be removed to give silk-like fibers. The technology to reduce the weight of fabrics and mixed spinning technology of fibers that shrink differently are the basic technologies.

7.2.4.2 Production of fabrics using a new concept during the 1980s

Natural fibers are limited in the type of processing which they can be subjected to. Synthetic fibers, on the other hand, have no such limitation in molecular planning and processing. Inevitably, therefore, during the 1980s, research was concentrated on developing new synthetic fibers with a new aesthetic sense and functionality that could not be provided by the normal fibers, using the accumulated technologies during the past decades as already described. Thus came *Shin-gosen*, which does not refer to a specific material such as nylon or polyester. It is a new category of fiber. Some of the basic technologies that have been built upon to produce *Shin-gosen*, particularly conjugated or mixed spinning combined with surface treatment, are shown in Table 7.4.

Table 7.4. Examples of the products of conjugated or mixed spinning combined with surface treatment technology

Trade mark	Producer	Fiber	Cross-section	Cross-section, surface after finishing	Speciality of products
Treview	Kanebo Ltd	Random conjugated finishing			Dry, spun-like natural feeling
Fontana	Asahi Chemical Industry	Mixed or conjugated spinning		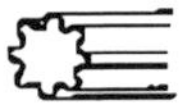	Dry-spun silk-like natural feeling
SN 2000	Kuraray Co.	Inorganic particle mixing			Dry-hand, deep colour
Sillook Royal	Toray Industries Inc.	Radial conjugated spinning		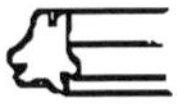	Bulkiness, silk sound
Sillook chatelaine	Toray Industry Co.	Inorganic particle mixing			Dry hand, rayon-like
Louvro	Toyobo Ltd	Inorganic particle mixing			Dry-hand, rayon-like, higher bending stiffness
Rapitus	Teijin Ltd	Copolymer or mixed spinning			Natural feeling, crispy, spun-silk-like

7.2.5 *Feminine* Shin-gosen

7.2.5.1 High shrinkable levels

Mixed spinning of fibers with different levels of shrinkage yields fabrics with fullness. With the appearance of super-high-shrinkable fibers and self-elongation fibers, a highly improved fullness was possible. The previously high-shrinkable polyester fibers had shrinkage levels of 15% in boiling water. However, super-high-shrinkable fibers give 30–50% shrinkage levels. Extremely bulky material can be given by the mixed spinning of fibers with highly different shrinkable levels as illustrated in Fig. 7.6. Self-elongation fibers are obtained by drawing and shrinking the fibers at a lower temperature to give a low crystallinity. Then treating the fibers at a higher temperature during the dyeing process will crystallise and elongate the resulting loops on the surface of fabrics and give an additional fullness to the fabrics. The stages are illustrated in Fig. 7. 7.

7.2.5.2 Strongly twisted composite fibers as basic for the production of new worsted

Composite fibers are bulky fibers provided by mixed spinning of filaments with different levels of elongation, and then false twisting. The fibers thus

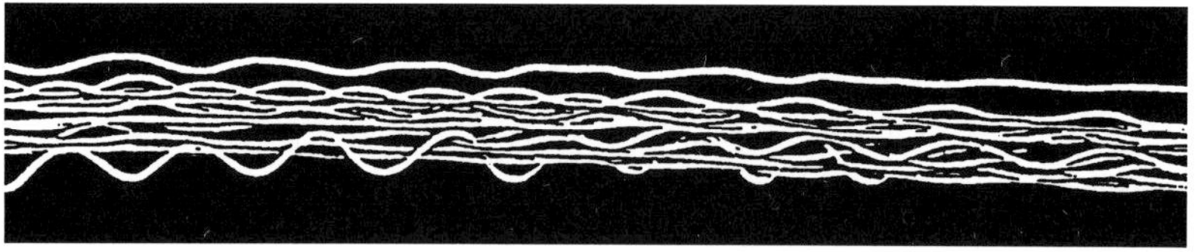

7.6 Ultra-bulky material made of mixed spun fibers with different shrinkage levels.

obtained have complicated crimps, and therefore provide a highly wool-like appearance with better fullness compared with the ordinary fibers. Worsted produced by the *Shin-gosen* concept produces awareness of high quality, and has good drapability, with improved fullness, because the crimps in the fiber are of complex design. Such fullness and excellent hand feel cannot be

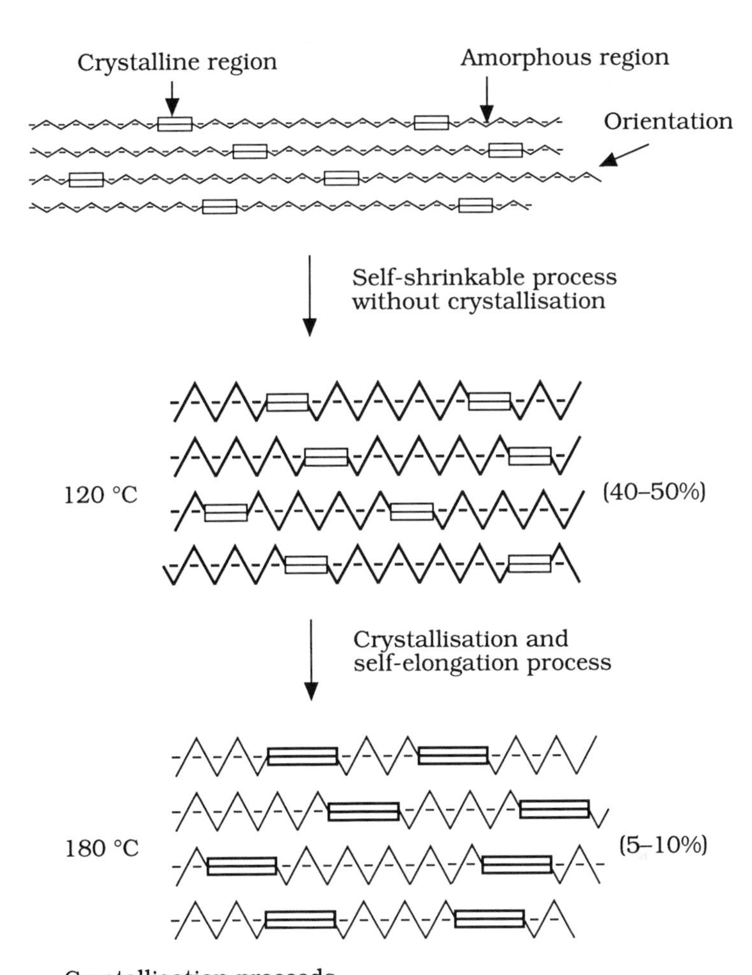

7.7 Self-elongation fibers by temperature change.

produced using conventional fibers because the crimps of necessity have to be simple.

7.2.5.3 Preparation for the weaving and dyeing processes

Having used strongly twisted fabrics to give good drapability, the loose filaments must be woven in a way that would not have been considered previously, so that during the weaving the new characteristics of the filaments are not lost. Designing of the dyeing process thereafter is an important step to bring out the properties of the fibers.

7.2.5.4 *Extension of variation of hand touch*

A new technology was also needed to introduce a variation in hand touch of *Shin-gosen.* In addition to mixed spinning of fibers with different levels of shrinkage and strongly twisted composite fibers, new rayon-like fabrics with a dry touch and high drapability could be produced by blending fine inorganic particles into polymer in the mixed-spinning process. One of the objectives of the mixed-spinning is to increase the specific gravity of the fiber and to improve the drapability, which the inorganic particles can achieve because they have a high specific gravity. Generally, the mixing is carried out in the polymerisation process, but it can also be done in other stages. The content of inorganic particles should be less than 10% by weight to avoid damage during processing. Extremely fine fibers are also incorporated to give a fine touch to the fabrics. Such fibers can be used in combination with other fibers or not, according to the requirement of the consumers for the type of hand feel.

The share of *Shin-gosen* is about 50% for new silky fabrics (mixed spinning of fibers of different shrinkage), 30% for new worsted (composite fibers), 20%

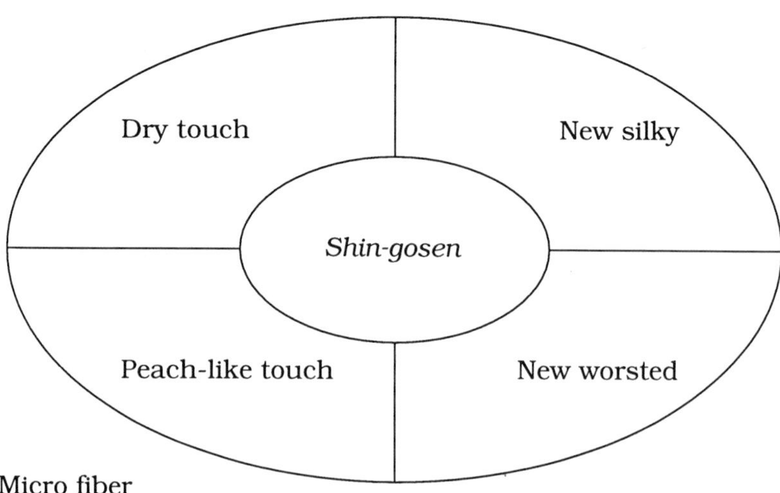

7.8 Classification of technologies to impart the various types of hand feel to *Shin-gosen.*

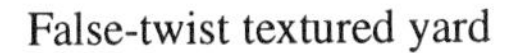

False-twist textured yard | Complexed yarn | Complexed yarn (slub)

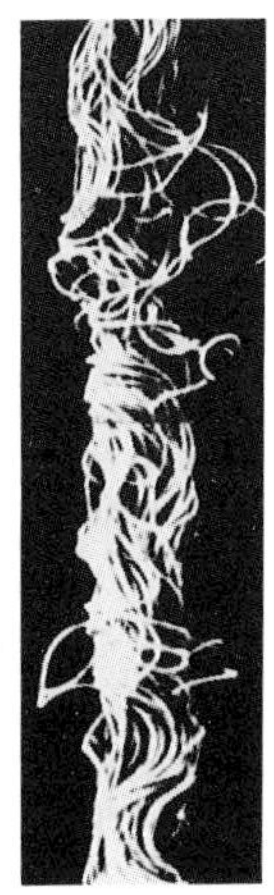

Crimp effect

Twist effect – fluff effect
Space among fibers
Colour mixable

Highly twist effect

Wool-like hand

Linen-like hand

7.9 Complexed textured yarn.

for new-rayon fabrics (dry-to-touch) and 10% for fabrics using ultra-fine fibers (as used for the production of artificial suede). The classification of technologies to impart the types of hand feel of *Shin-gosen* is shown in Fig. 7.8. Generally, the term "microfiber" is used in Japan for artificial suede. *Shin-gosen*, in this regard, may be considered as fabrics made by the use of microcraters. Figure 7.9 illustrates the various methods used for textured yarns to achieve variations in hand-feel.

7.2.6 *Casual (comfortable)* Shin-gosen

"Comfortable" *Shin-gosen* arose with the change of fashion trend and consumer pressure. The development went along two paths: one is the development of water/sweat-absorbent or easy care materials, and the other is

Table 7.5 Methods to produce "comfortable" *Shin-gosen* materials

Basic technology				Materials	
Cross-section	Added particles	Polymer	Fiber processing	Name	Producer
Circular	Ceramics	Regular		Alteene Estmoule Xye	Toray Teijin Kuraray
Micro-slit			Differently shrinkable	Ceo	Toray
Cross	Ceramics		Thick and thin	Space-master	Kuraray
Circular, hollow			Spinning	Aero-capsule	Teijin
Triangular, hollow	Ceramics		Thick and thin	Gulk	Asahi Chemical Industry

Table 7.6 Methods to prepare water/sweat-absorbent and easy-care materials

	Basic technology			Materials	
Cross-section	Added particles	Polymer	Processing of fibers	Name	Producer
Circular, hollow				Wellkey	Teijin
Circular, hollow			Conjugate Thick and thin	Aege	Mitsubishi Rayon
Circular, hollow		Dissolution		Kilatt P	Kanebo

the development of composite fiber materials with a new hand touch. The basic technologies are summarised in Table 7.5.

Since polyester is hydrophobic, it lacks moisture and water absorbency. Therefore, efforts to improve its hand-feel and other functionalities, such as water absorbency and easy care, must begin at conception. This started first with polymer modification and processing, then changes in cross-section, surface modification and structure. However, all these needed to be combined to give water/sweat absorbency and easy care, as described in Table 7.6.

7.2.7 "Comfortable" Shin-gosen *using composite materials*

The movement to develop a specially "comfortable" *Shin-gosen* using composite materials started with the polyester manufacturing companies at the end of 1992. Composite materials include not only the mixed spinning of wool or rayon with polyester, but also composite materials made up of

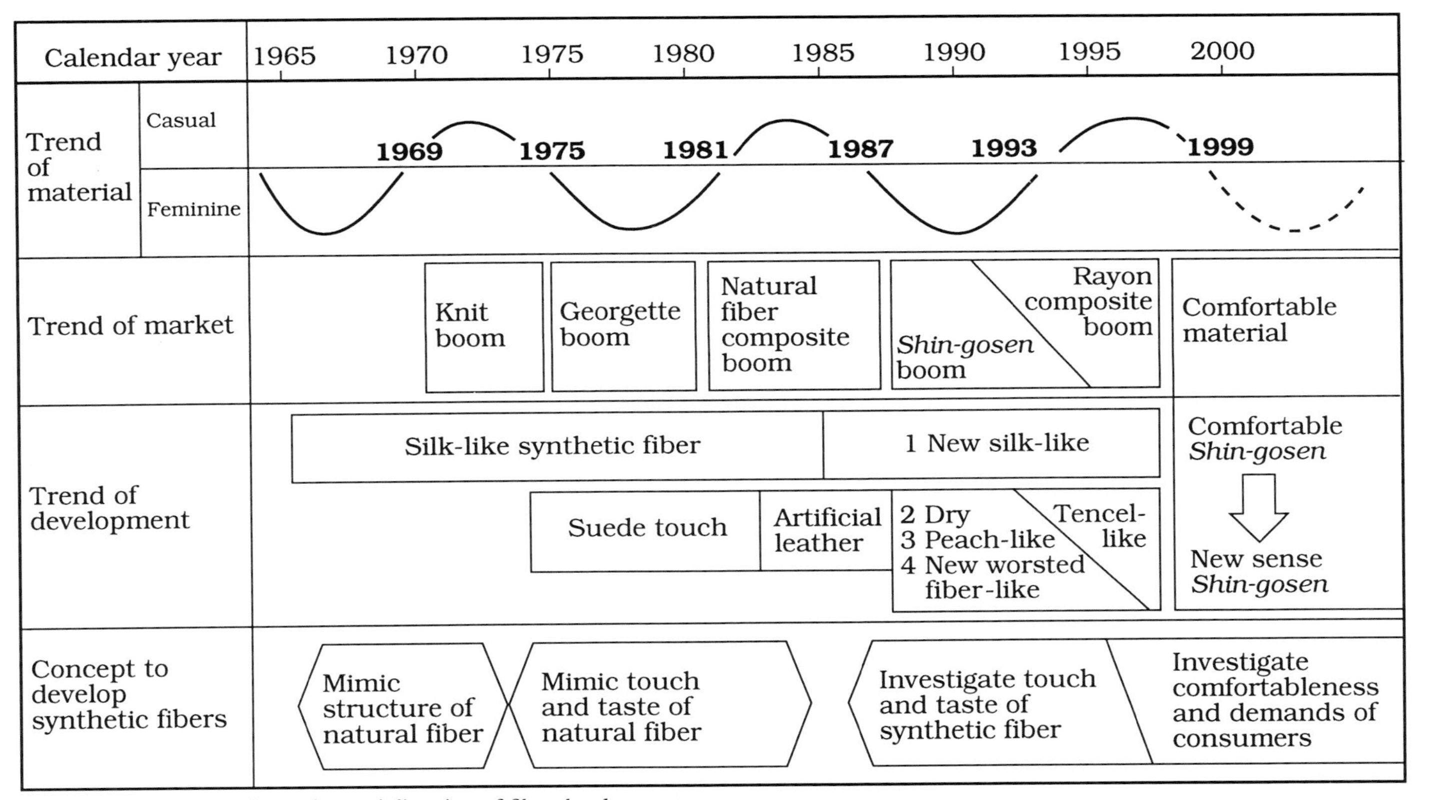

7.10 Change in market and direction of fiber development.

Table 7.7 Comfortable *Shin-gosen* made of composite materials

Basic technology	Content
1 Fancy fabrics	Express rough texture of fabrics using *Shin-gosen* such as new rayon-like, new-worsted yarn
2 Finishing	Express natural character by uneven dyeing, wrinkle or surface finishing
3 Composite spun-like	Expressed by twisting or entangling ultra-fine fibers around thick filament in complex and uneven way
4 Spun composite	Expressed by mixed weaving of *Shin-gosen* and spun yarn Expressed by filament/staple fiber composite materials using *Shin-gosen* and spun yarn conjugate-spun-and-twisted Conjugate-spun-and-twisted: core-sheath type, side-by-side type, and uniformly mixed woven type
5 Rayon composite	Expressed by mixed weaving of *Shin-gosen* and rayon Expressed by *Shin-gosen* and rayon conjugate-spun and twisted
6 Natural fiber composite	Expressed by mixed weaving of *Shin-gosen* and natural fibers (cotton, wool) Expressed by *Shin-gosen* and natural fibers (cotton, wool) conjugate-spun-and-twisted

different fibers. The purpose of the such new composite materials is to provide together good hand touch, appearance and functions not possible using a single material. Such "casual" *Shin-gosen* is distinguished from feminine *Shin-gosen* by using the title "comfortable *Shin-gosen*". The change of the market and the trend of material development thereafter is shown in Fig. 7.10.

Comfortable *Shin-gosen* is again produced by a combination of technologies, for example, composite fibrication, composite spinning and composite dyeing to achieve a more natural feel and appearance (see Table 7.7). Composite-spun-like fibers, spun filament staple composites and rayon composite fibers are the main materials for such fabrics with rough and natural character. The material should be comfortable (light, elastic, easy to wear, good for tailoring and good recovery of shape) to produce clothing. *Shin-gosen* of this type also needs to have in-built moisture absorbency and release. It will require more development to improve the basic chemical properties and multi-layer structure of the base materials. Therefore, *Shin-gosen* at present is integrated with other naturally derived materials such rayon, cotton and wool to provide the degree of comfortableness required.

7.3 Design of specialist fibers

Specialist fibers, requested by market trends, have been developed by methods outlined in Fig. 7.11. Properties required for commodities are often opposite from fibers, for example, with a different combination of soft and

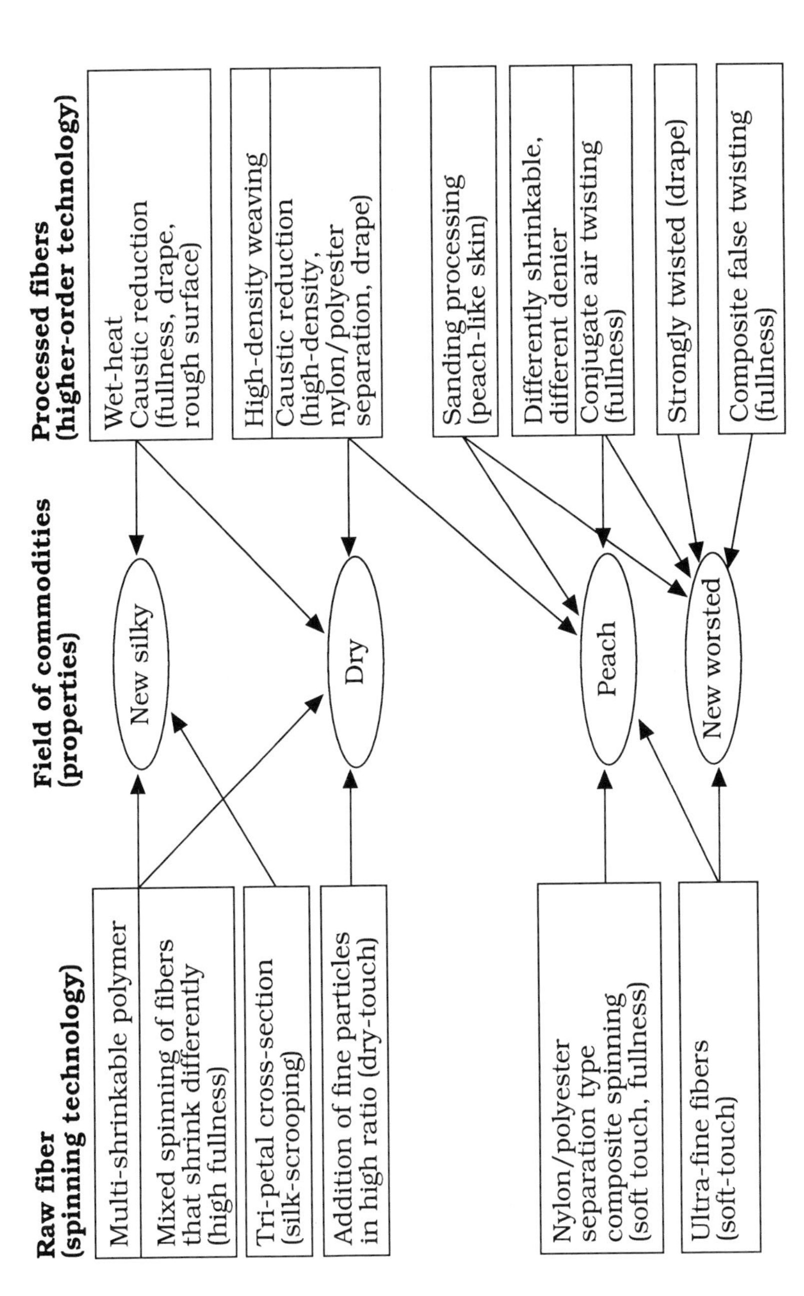

7.11 Necessary properties for specialist fibers.

spread-and-anti drape/stiffness, drapability and repellency, soft and dry-touch. To fulfil such demands, high-level technologies are necessary to retain the superior basic properties without damage by the processing methods. For this type of development, new polymers could be needed, which are now in development. False twisting and fluid entanglement are processes that can convert such polymers with special properties into fibers for commodity use.

7.3.1 New ("sense") Shin-gosen

A fashion started in 1996 was to achieve a polished casual trend. Predictions are that in the future, modern elegance will be required even for casual wear. Materials for very elegant outwear, with good-looking surface touch, lightness, comfort, with attractive taste, supported by good tailoring will be required when developing this *Shin-gosen*. The material should be light, easy to wear, with dry-hand touch, tension, tenacity and water repellency. To this end, Toray has developed "Cheddy", "Cheddyleshe" and "Cheddyelve" by applying recently developed higher processing technology. "Cheddy" was developed by combining a spinning technique to control the crystallisation and orientation of ultra-thin fiber and then combining ultra-thin with ultra-thick fibers (Fig. 7.12). "Cheddyleshe" uses a new fiber with an air hole structure and ultra-fine hollow fiber. It is a multi-dry material, which is light,

7.12 Cheddy: the fiber that combines ultra-fine and ultra-thick fibers.

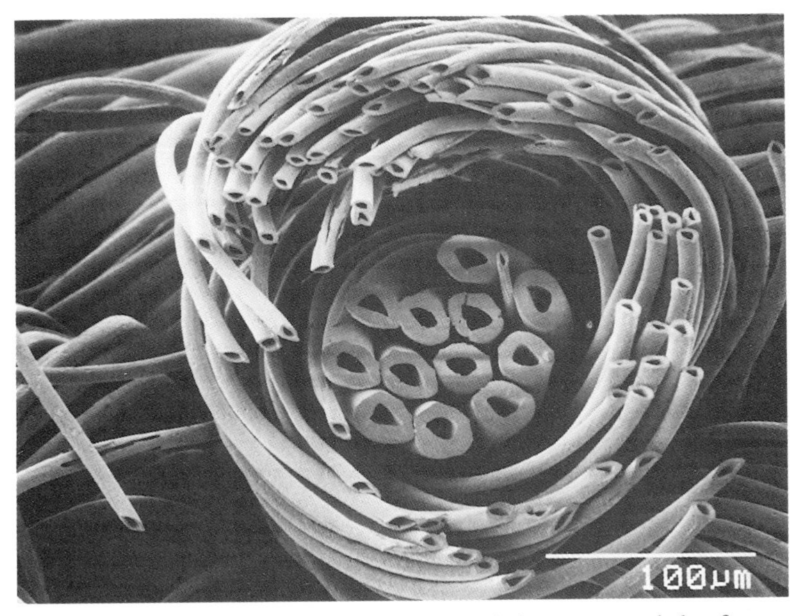

7.13 Cheddyleshe: the fabric with an air hole structure and ultra-fine hollow fiber.

7.14 Cheddyelve: the fiber with a gentle touch with a micro-random wave structure.

comfortable to wear, mild and dry, and highly water repellent. By applying a "cyclone structure" to the fiber, twinning fine fibers on ultra-thick fiber, a multi-layered porous structural material is obtained (See Fig. 7.13). "Cheddyelve" was developed to attain a micro-random wave structure, by controlling the orientation and morphology of ultra-fine fibers by applying a newly developed "highly twined structure stabilising technique" (Fig. 7.14). As the result, fibers with finer and gentle touch were developed. They can be used for outerwear, such as skirts, trousers, dresses and coats for women and outerwear, trousers and wind-breakers for men.

For products that have a higher "sense" and function than *Shin-gosen*, the term "*New Shin-gosen*" was introduced around 1993/94. While retaining the basic "sense" of *Shin-gosen* (silkiness, worsted-like, dry and peach-like touch, etc.) and functions (sweat absorbency, electrostatic-proof, stretch-ability, etc) the *New Shin-gosen* includes also lightness, warmth, a cool touch and moisture permeability. It will penetrate the market from now on.

More recently "funny fibers" or very interesting fibers became topics of development. These have sometimes practical, but often impractical, properties such as thermochromic, photochromic, perfumed, antibacterial, deodorant, heat-storing, water-repellent, electroconductive effects (see also Chapter 3). These have also benefited from the new technologies developed recently, particularly various spinning technologies. Examples are conjugate fibers made from a terpene-containing polymer and a fiber-forming polymer; antibacterial fibers in which zeolite particles containing silver are blended; light-to-heat transferring fibers in which carbonised zirconium particles are blended, and water-absorbent fibers having a hollow and porous structure or a slit as a water passage.

Mention should be made also of K-II developed by Kuraray. By any standard, it must be regarded as a dream fiber, since there does not appear to be anything comparable. It is soluble in water, it has high strength and has significant thermal resistance. Toray and Komatsu Seiren have now developed a polyester fiber that absorbs moisture and heat very quickly and releases them to the atmosphere. To produce this, fiber polyester is grafted with a polymer produced from a hydrophilic vinyl monomer together with a protein component. These examples show clearly that we can further expect revolutionary new fibers in the near future.

7.4 Fabrics for relaxation using 1/*f* fluctuations

Nisshinbo has developed a range of new products that have a hand-crafted feeling through the application of 1/*f* fluctuations to the spinning, weaving or knitting processes. Such fluctuations are found widely in nature, as in a breath

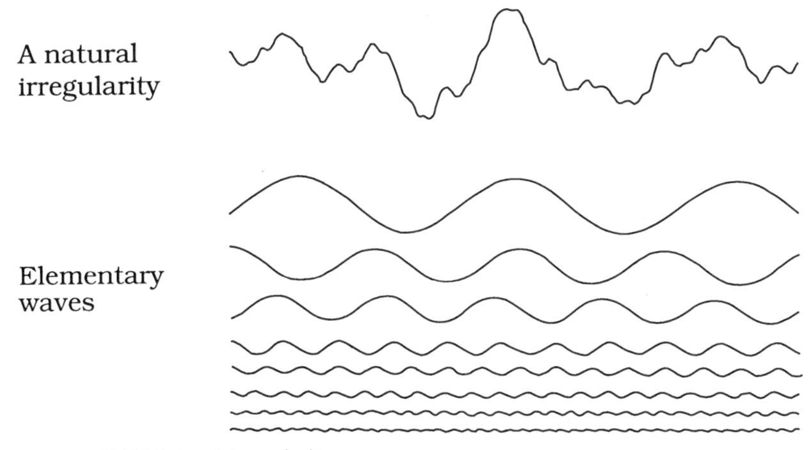

7.15 Natural irregularity.

of wind or the murmur of a brook; these create peaceful feelings. The idea of giving irregularity to industrial products is very new and revolutionary because, up to now, these products have been produced to be as uniform as possible.

$1/f$ fluctuations may be called a rhythm in nature; they can be observed widely in natural phenomena and give us a feeling of relaxation. In general, natural phenomena or natural materials have irregularities that are sometimes pleasant or unpleasant to us according to their state. A typical natural irregularity is illustrated in Fig. 7.15 and looks like an irregular wave, having no rules.

However, by analysing such irregularities, it can be established that they are the result of a combination of other simple elementary waves. Then, when we observe the size of these elementary waves against their frequencies, certain natural irregularities show an inverse proportion between the wave size and frequency. Such irregularities are called $1/f$ fluctuations. By this analysis method, it has been shown that there are many $1/f$ fluctuations in natural phenomena and these phenomena give us relaxation or impressions of beauty. The observation that such $1/f$ fluctuations are encountered in beautiful music led to the conclusion that these are not only a universal rhythm in nature but also closely connected to comfort and beauty.

It is then considered that the universal rhythm in nature are also present in the human body. Then human beings recognise a resonance to such a $1/f$ fluctuation in some phenomenon they feel a familiarity with that phenomenon. Therefore, we become relaxed when we recognise a breath of wind, hear the sound of waves or the murmur of a brook or see the twinkling of the stars.

7.16 Application of $1/f$ fluctuations to fabrics.

Nisshinbo applied the concept of $1/f$ fluctuations to yarns and textiles, and has developed a spun yarn with a $1/f$ fluctuation through the use of a special spinning system which can control the degree of draft. This yarn is made mechanically, but it has a hand-spun appearance.

This yarn is now being applied to many textile products such as handkerchiefs, curtains and denim. These products are not uniform in their appearance and are designed to give us a feeling of relaxation. Until now, uniform appearance has been the most important quality standard in industrial production. Therefore, the idea of producing industrially uneven products of natural irregularity is quite new (see Fig. 7.16).

7.5 Some new arrivals

7.5.1 Super Elequil

Mitsubishi Rayon has developed this sheath–core type electrical conducting acrylic fiber. It is the first sheath–core type acrylic fiber employing the composite-spinning method. Electric conducting particles, 0.2–0.3 μm in diameter, are ceramics coated with metals and inserted into the core. It prevents static electricity, thus reducing dirt and dust particles, and also prevents the fabric from clinging. The antistatic properties are maintained even after repeated wearing and washing, and now provide a new element of safety for working clothes (Table 7.8).

Table 7.8 Standards of safety achieved by Super Elequil

Sample	Charge quality Sweater	Knitted fabrics
Test method	JIS T 8118	JIS L 1094-C
Standard	<0.6 μC/m^2 product	<7.0 μC/m^2
Super Elequil blended fiber	0.23	3.68
Regular acrylic fiber	0.81	8.70

Sample: Acrylic fiber/wool (70/30), 2P14G rib knit stitch.
Standard: Japanese Labor Dept. Standard test for antistatic working clothing.

7.5.2 "Clean Guard": a deodorising fabric

Komatsu Seiren has developed "Clean Guard" fabric to kill various odours encountered in daily life. Deodorising agents introduced into the fabric during finishing eliminate odours by chemical reaction and physical adsorption. It is particularly effective against ammonia, trimethylamine, methylmercaptan and hydrogen sulphide. In addition "Clean Guard" suppresses sweat and sock odours (due to isovalelic acid) and body odour (due to pelargonic acid, capric acid, caproric acid, etc.). Polyester fabrics will mainly be used for "Clean Guard", since Komatsu Seiren's speciality is dyeing and finishing man-made fibers, particularly polyester based.

7.5.3 New water-soluble PVA Fibers

Water-soluble PVA fibers have been prepared by a new cooled gel wet-spinning process (See Fig. 7.17). They have the following properties:

1 Water solubility between 5 and 90 °C.
2 Non-woven making processability with heat calendering and embossing.
3 Heat sealability.
4 Storage, stability and easy handling under high humidity condition.
5 Salts and plasticiser free.

In the method, both the dope solvent and solidifying liquid solvent are organic solvents. Accordingly, various raw materials like very low saponification degree can be used in this process. By selecting a particular saponification degree of PVA, the water-dissolving temperature of fibers can be controlled. These fibers that dissolve in water at temperature lower than 80 °C, have good dimensional stability under high humidity conditions and do not contain any salt and plasticiser.

Conventional PVA fibers do not have a heat calendering property. In order to solve this problem, a new method was required, i.e. making a blend fiber

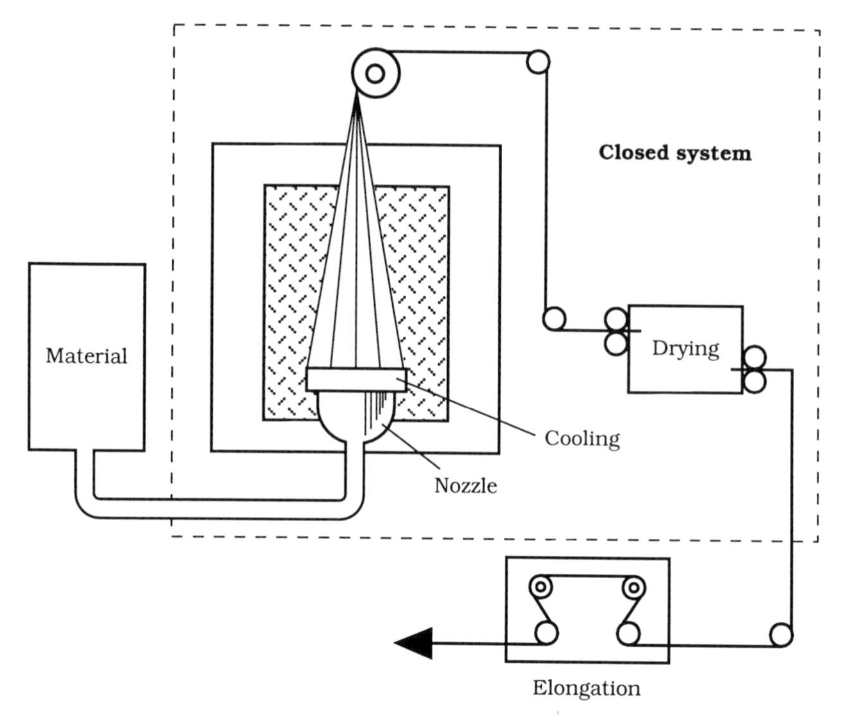

7.17 Cooled gel wet-spinning process of K-II.

that consists of high melting point PVA and very low melting point PVA which is highly adhesive. By subjecting the blend polymer to a cooled gel-spinning method, an island–sea structured fiber was obtained. The sea and island components are respectively high melting PVA and low melting PVA. When the webs of this blend fiber are processed with heat calendering, they become bonded. The adhesive mechanism is believed to involve the heated sea component being crushed with the pressure and its melted island component adheres to its sea component or the island component to each other. It is a thermocompression bonding fiber that needs both heat and pressure for adhesion.

No water is used through the whole process. The products are water soluble at a low temperature range and have thermocompression bonding properties. It is quite new for non-wovens to be produced that completely dissolve even in 20 °C water. These new water-soluble PVA fibers can be used for dry laid non-wovens, and are very useful for chemical embroidery lace backings, water-soluble packaging materials, plant growth aid materials, and medical, sanitary, industrial and household applications.

8 Cellulosic fibers

Since Count Chardonnet first developed rayon as a silk-like fiber, there have been many developments in the field of synthetic cellulosic fibers that give cellulose, once again, an opportunity to challenge the synthetic fibers in their applications and environmental effects. At the beginning of the 20th century the world demand for textile fibers was 3.9 million tonnes, made up almost entirely of cotton (3.2 million) and wool (0.8 million). By 1950 the total demand had risen to 9.4 million tonnes of which 6.6 tonnes was cotton, 1.1 million wool and 1.6 million tonnes the synthetic cellulosics. In 1980 the production of synthetic cellulosics peaked at 3.6 million tonnes, but declined to 2.8 million tonnes by 1993. This drop can mainly be attributed to the collapse in demand in Eastern Europe following the political changes there.

Meanwhile the world population over the same period increased from 1.6 billion to 4.04 billion. It is clear that over this time the cellulosics could not provide the same flexibility that the synthetics had achieved. During the same 50 year period the demand for synthetic fibers had reached 18.6 million tonnes with cotton accounting for 18.5 million tonnes in consumption. The main reasons for this lack of competitivity was the inability of the synthetic cellulosics to achieve a built-in functionality; they could not match the variety of applications of the synthetics or their price. Whereas the price of cotton dropped because of the increase in world production, and the synthetic fibers' prices fell because of over-capacity and lower production costs, the price of the cellulosics continued to increase. This was mainly due to the greatly increased costs associated with pollution control associated with the old chemical pulping processes.

The world population is growing fastest in the regions of the world with a warm climate. It is likely that by the year 2000 the total world textile requirement will be of the order of 50 million tonnes. If the present 56% demand for absorbent fibers is maintained, as is likely, then there will be a call for 28 million tonnes of absorbent fibers. The production of this category at

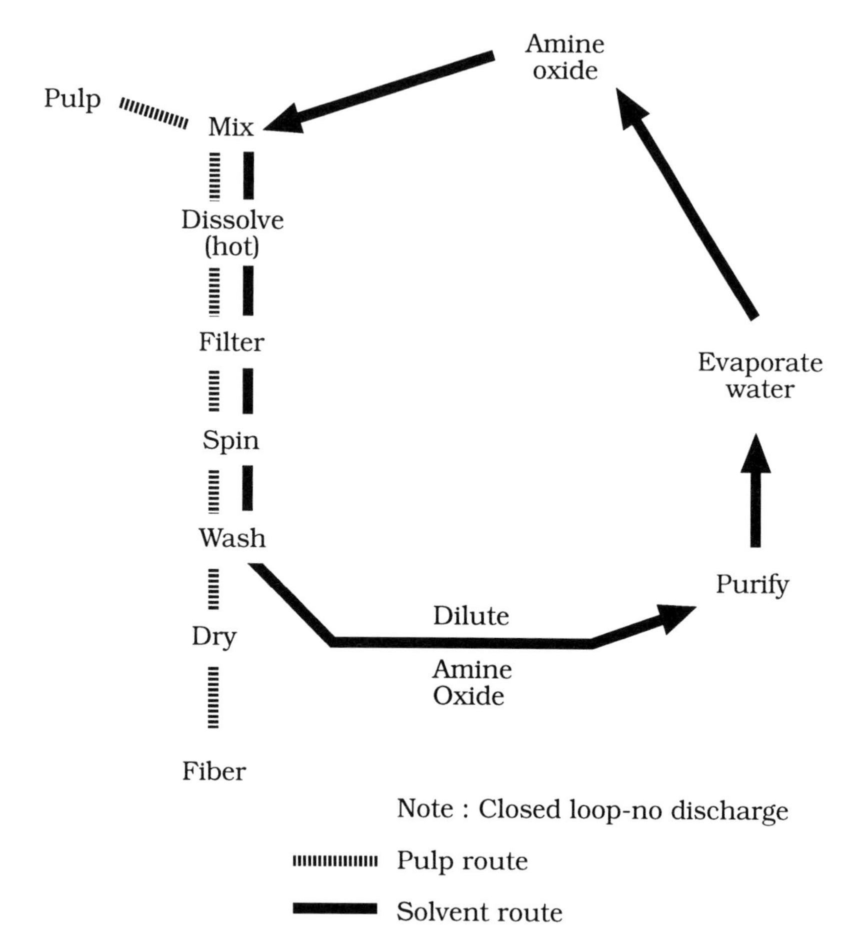

8.1 Outline of the Tencel process.

present is 18.6 cotton, 1.6 wool, 2.8 cellulosics and 1.6 viscose staple, which is 24.6 million tonnes in all. It is unlikely that cotton production can be greatly increased because the area for planting has remained stable for the last 40 years. The need to use land to produce food has taken precedence. Undoubtedly there is an opportunity for the cellulosics, and it is for this reason that innovative developments have taken place recently to overcome the shortcomings of the past.

8.1 New solvent systems

If processability is to be improved for the cellulosics, then new solvent systems are needed. Research has yielded many which are promising:

paraformaldehyde–dimethyl sulphoxide cyclic amine oxides; nitrosyl chloride dimethyl formamide; chloral dimethyl formamide, for example. The two that have won through into actual commercial processing are sodium hydroxide and *N*-methyl morpholine *N*-oxide (NNMO).

8.1.1 Courtaulds' Tencel, Lenzing's Lyocell and Akzo Nobel's NewCell

These new fibers are synthetic cellulosic fibers produced from morpholine *N*-oxide solvent systems. Tencel is the new cellulosic fiber developed by Courtaulds, one of the pioneer companies in rayon production. It is produced from natural cellulose in wood pulp using a special technique to spin from NNMO solution. It is an environmentally clean process with virtually total recycling of the solvent as shown in Fig. 8.1. The project code-named "Genesis" started on 1981 and a semicommercial plant was built in Grimsby, UK in 1988. A full-scale production facility was commissioned in Alabama, USA, in 1992. The basic patent is entitled "Process for Shaped Cellulose Article prepared from a solution containing Cellulose Dissolved in a Tertiary Amine N-Oxide Solvent", and the inventor was Clarence C. McCorsley, III. The patent was filed on 2 March 1989 and was assigned to Akzona Inc. Asheville, now Akzo Nobel. It gave the licence to produce staple fibers to Lenzing, Austria, in 1988 and to Courtaulds, UK in 1990.

The various terms and developments with this new fiber can be confusing, so identification of the progress made is important:

1 "Lyocell" is a synthetic cellulose fiber produced by spinning from a solvent made of wood pulp in amine oxide.
2 "Tencel" is the first commercialised Lyocell (staple) fiber by Courtaulds, UK.
3 Courtaulds constructed its semi-commercialised plant in the UK in 1988, and went into full production via its commercial plant in the USA in 1992.
4 Lenzing took a licence on the basic patents in 1988, and since 1990 has been running a pilot plant, with a view to going into commercial production in 1998. Its staple fiber is termed "Lyocell".
5 Courtaulds and Akzo Nobel announced a feasibility study of Lyocell filament yarn which was completed in 1996. The filament yarn will be termed "NewCell".

Courtaulds were, therefore, the first to establish a production process and has sold all over the world with great success. Production of Tencel at the end of 1998 will be 90,000 tonnes per year.

Table. 8.1. Fiber properties of Tencel compared with other fibers

	Tencel	Modal	Viscose	Cotton*	Polyester**
Titre(dtex)	1.5	1.7	1.7		1.7
Tenacity (cN/tex)	40–42	34–36	22–26	20–24	55–6
Elongation (%)	14–16	13–15	20–25	7–9	25–30
Wet tenacity (cN/tex)	34–38	19–21	10–15	26–30	54–58
Wet elongation (%)	16–18	13–15	25–30	12–14	25–30
Tenacity (@ 10% ext)	35	23	16		26
Wet modulus (@ 5% ext)	270	110	50	100	210
Moisture regain (%)	11.2	12.5	13	8	0.5
Water imbibition (%)	65	75	90	50	3

* US middling.
** High tenacity.

Lyocell long fibers are being produced in a pilot plant in Germany, in cooperation between Akzo and Courtaulds. Akzo provide the technique to produce long fibers, with Courtaulds inputing the production technology developed for Tencel. Early 1998 saw the start of production in a semi-commercial plant. Marketing and application is being introduced under the leadership of Akzo. The great attraction of the Lyocell type of fiber is its considerably greater strength and versatility compared with rayon and other cellulosics. Its production process also is environmentally sound. It is Tencel that has taken the lead, and currently Lenzing's Lyocell staple (short) fiber has only a third of the production capacity compared with Tencel. The main attention will, therefore be given to Tencel.

Tencel retains all the natural properties of a cellulosic, with good moisture absorbency, comfort, lustre and biodegradability, with coloration characteristics similar to rayon. The properties of Tencel are summarised in Table 8.1. It has high strength, with only 15% loss of strength in the wet state. Exceptional wet modulus results in very low fabric shrinkages. Tencel has a round cross-section and a good open fiber appearance for ease of subsequent processing. It is an ideal fiber for blending with other fibers to give very strong yarn, even at low blend levels.

Tencel produces apparel fabrics with good aesthetic versatility. It satisfies a wide range of fabric needs from fine fashion ladies' lightweight blouse- and dress-wear through moderately heavy skirting and suitings. Its versatility also extends from active leisure wear through to performance fabrics, denims, work-wear and industrial apparel. Tencel produces more luxurious drape effects than cotton. Either alone or in blends with other fibers it provides a variety of handling effects, allowing fabric finishers to demonstrate their skills. Like rayon, Tencel has a very efficient dye uptake and provides natural, bright, vibrant colours. It has a dyeing compatibility with other rayon fibers, which allows a wide range of options for substrate fabrics.

The high strength, rigidity and wet modulus of Tencel translate through fabric structures that show exceptional strength, especially when wet, and very low shrinkage. Additionally, the cellulosic character provides very good thermal stability and low creep. These good physical characteristics allow a broad range of technical applications, particularly spun industrials, disposables and durables. Additionally the environmentally favourable nature of the manufacturing process, combined with comfort levels and biodegradability, give this new fiber a good opportunity to carry the fight back to the synthetic competitor. The development offers an opportunity of having the processing advantages of the synthetics combined with the benefits of the natural fiber.

Although the basic technology to produce Tencel and Lyocell staple fibers is the same, there are subtle differences. Lyocell was improved in its degree of crystallinity and orientation and targeted for use in clothes. Tencel is washed and purified directly after spinning as bundles of long fibers, whereas Lyocell is washed after cutting into short fibers. The treatment gives fibers of different strength. The tensile strengths are:

Lyocell 4.0–4.5 g/denier (Dry); 3.5–3.0 g/denier (wet)

Tencel 4.5–4.8 g/denier (Dry); 4.0–4.3 g/denier (wet)

The Young's Moduli are:

Lyocell 900 kg/mm^2

Tencel 1300 kg/mm^2

Consequently, Tencel finds application not only for clothes but also for industrial materials.

NewCell, as noted, is the name given for the first Lyocell filament yarn being developed by Akzo. It is meant to complement the spun yarns produced from Tencel staple fiber. A NewCell pilot plant has been erected at Obernburg, Germany, to carry out process engineering in preparation for upscaling to a production plant. As with the other fibers produced by NMMO technology, NewCell is able to offer the same advantages as the existing cellulosics, such as wear comfort, moisture absorption and versatility in applications. Its characteristics are:

1 It can be spun in very fine total deniers, and so can find new uses as it is a cellulosic microfilament yarn.
2 It has a tenacity in the dry state that is twice that of existing cellulosic filament yarns.
3 It has excellent dimensional stability in the fabric and the garment due to the low non-continuing shrinkage of the yarn and its reduced water retention.

4 It can be texturised and it has the ability to fibrillate.
5 Garments made of NewCell are machine washable, but unfortunately still have to be ironed.

There is still some way to go in market development and it will be approaching the next century before the full potential can be elucidated.

Lenzing AG is putting its faith in Lyocell fiber and decided in May 1995 to install a 20,000 tonne per year production line which will go into operation step-by-step in 1998. Many technological hurdles needed to be overcome. Each of the companies is attempting to place its particular product in a market sector where its properties can be exploited. Lenzing emphasises the softness of Lyocell, which is not equalled by cotton, even when treated with enzymes and softeners. The dyeing behaviour too is comparable to other cellulose fibers. The basic high strength of this fiber makes it possible to treat it at different stages in the textile chain in a dry and wet state. A range of handle and look variations can be produced, the most popular being those that resemble wool or silk. The nature of the fiber offers considerable innovative potential, and clearly the race is on to gain the best from these new fibers.

8.1.2 Sodium hydroxide as a solvent

Sodium hydroxide has proved an interesting and important new solvent for cellulose. The dissolution process, however, cannot readily be achieved without some pre-treatment of the raw material. To utilise wood cellulose, lignin and other non-cellulosic materials need to be separated from it. Delignification of wood is carried out by a sulphite and prehydrolysed sulphate process. The end-use of the pulp determines the degree of signification required. Usually it is necessary to remove at least 50%, of the lignin, which then results in removal of at least 50% of the hemicelluloses and 10–15% of the celluloses from the wood also.

Several methods have, therefore, been tried to open up the wood structure and so make the cellulose more available for solvent systems. Preference, of course, is given in a practical situation to using various types of dissolving pulps, such as softwood prehydrolysed and hardwood sulphite pulps which can be made soluble in 9% sodium hydroxide by a combination of mechanical shredding and enzymatic treatments. These pulp fibers originating from the cell wall form ether and hydrogen bonded systems and contain cellulose with different degrees of polymerisation. The final reactivity and aqueous alkaline solubility of cellulose depends a great deal on the pulping process.

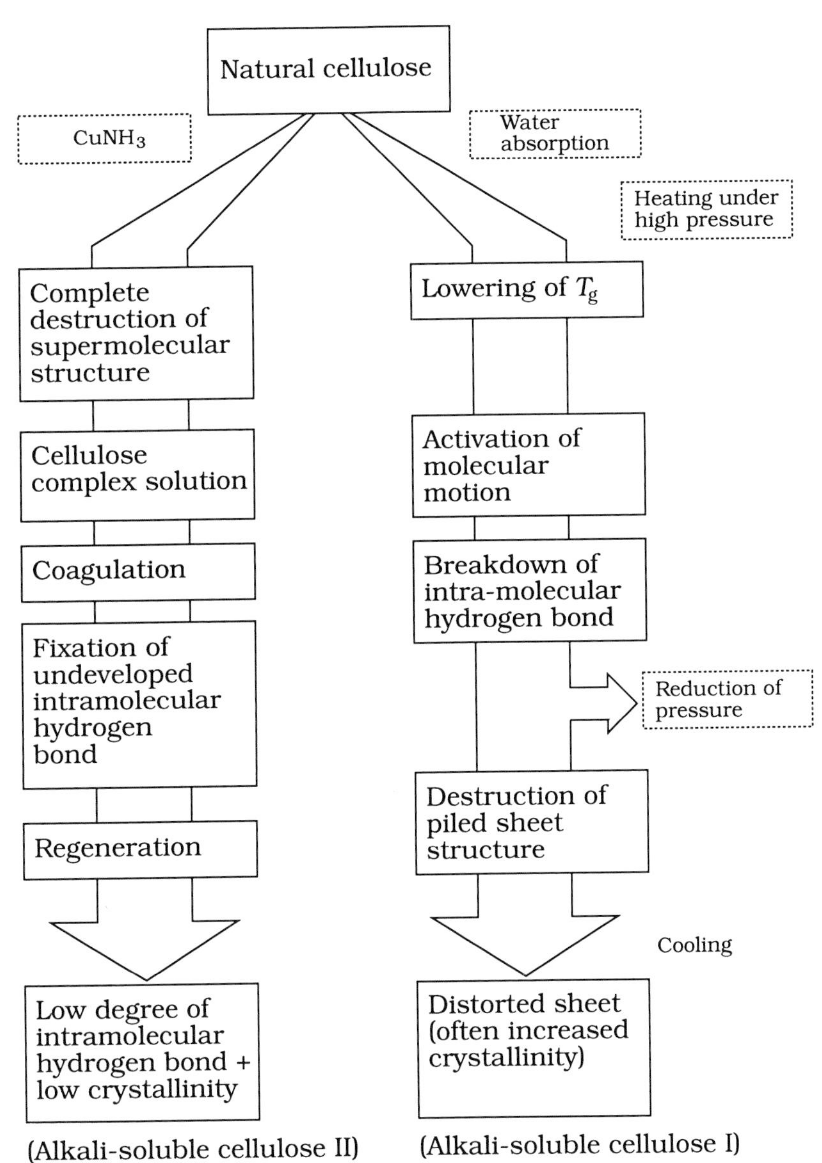

8.2 Underlying principle of the preparation of alkali-soluble cellulose (T_g = glass transition temperature).

Generally, the enzyme systems that degrade native cellulose require a combination of individual enzymes which either randomly or systematically from the chain ends hydrolyse the polymeric chains. Such a combination of enzymes from well-known *Aspergillus niger* and *Trichoderma reesei* have been used for the cellulose activation in the Celsol method of directly solubilising pulp in 9% alkali. However, the most successful method to open up the structure to solvent is that developed by Professor Kanji Kamide and co-workers at the Fundamental Research Laboratory of Fibers and Fiber-Forming Polymers at the Asahi Chemical Company in Takatsuki, Japan. These workers used steam explosion on a commercial scale, but only after establishing exactly what changes were induced in the cellulose to enable it to become wholly soluble in alkali.

The principle of preparing alkali soluble cellulose is shown in Fig. 8.2. First using the steam explosion process cellulose in the cellulose I crystal form can be prepared. The solubility towards aqueous alkali is governed by the degree of breakdown in the O—H——O intramolecular hydrogen bond. Once in solution, a whole new world opens up and another processable fiber with cotton-like properties can be produced. The fiber produced in this way by Asahi is a triumph for basic cellulosic research, and demonstrates that once the nature of a chemical process is understood, then the potential for commercial exploitation s is considerably greater.

8.2 New cellulosic fiber derivatives

The Institute of Chemical Fibers, Lódź, Poland, has developed an original technology to manufacture fibrous cellulose carbamate, which has a range of potential uses. The chemistry is very simple as illustrated in Fig. 8.3. It is important for fiber making that the cellulose carbamate be specially tailored. The average degree of polymerisation of the cellulose should be lowered to about 400 and adjusted so that the carbamate groups are evenly distributed along the cellulose chain. A research group at Nesteoy, Finland, achieved these two preconditions by employing a liquid ammonia treatment to activate its structure including a change in its crystallinity and enabling the even intercalation of urea into the cellulose structure. The group also used ionising radiation to reduce the degree of polymerisation. The Polish process did not use the liquid ammonia step, and was replaced by other chemical activators to modify the super-molecular structure of the cellulose.

A small pilot-industrial manufacturing plant has been set up in Zaaklady Chemiczne "Viscoplast" S A, Wrocław, Poland, and as a result substantial quantities of the product are available for trials, and practical applications such as fibers, film and other technical products. This technology presents many advantages. Several types of pulp are suitable for manufacture of cellulose carbamate. Good spinning solutions can be prepared. Solution of 9 wt% in sodium hydroxide can be produced which are extremely stable. Moreover, it can be well blended with viscose to prepare stable and good

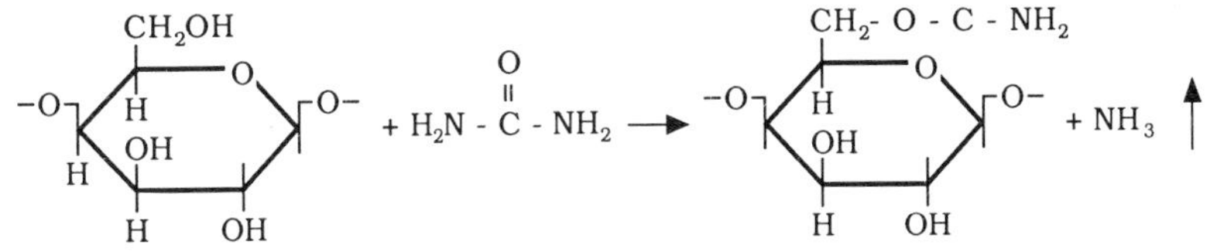

8.3 Scheme of reaction of cellulose with urea.

spinning solutions. The main objective for using the cellulose carbamate is to make fibers, which are spun by a wet method using coagulation as well as regeneration baths. The carbamate also has good film-forming behaviour. The major advantage is to allow this new process to replace the existing environmentally difficult viscose process in Poland. Most of the existing viscose fiber manufacturing equipment can still be used with the new technology. Moreover, cellulose carbamate is miscible with cellulose xanthate in the viscose process which enables a step-wise and safe introduction of the carbamate to the viscose process. It is of particular benefit to introduce greater water retention and produce high absorbent blended viscose fibers.

8.3 New environmental and cost saving developments

8.3.1 Electron processing technology

Chemical and physical changes are induced in cellulose when subjected to high-energy radiation. Degradation occurs due to chain scission and certain carbonyl and carboxyl groups are introduced into individual glucose units. The tertiary structure is also opened up, making the individual chains more accessible to solvents. Over the years such changes were regarded as harmful to the properties of cellulose, as for example cotton, which lost strength by this treatment. Recent work by Atomic Energy of Canada Ltd has demonstrated that the use of high-energy electrons, produced by an electron accelerator can induce changes in wood pulp which make it easier to process when used to prepare viscose. The changes are illustrated diagrammatically in Fig. 8.4. This new technology has considerable cost and environmental benefits and is now being considered by the major viscose manufacturers either for introducing on-line in their own viscose process or by having their pulp pre-irradiated at a central facility.

The changes and benefits that the electron process can introduce in the viscose process are illustrated in Fig. 8.5. Owing to the greater accessibility introduced by the radiation, the concentration of sodium hydroxide necessary can be reduced from 19% to 16%. The ageing step can be eliminated

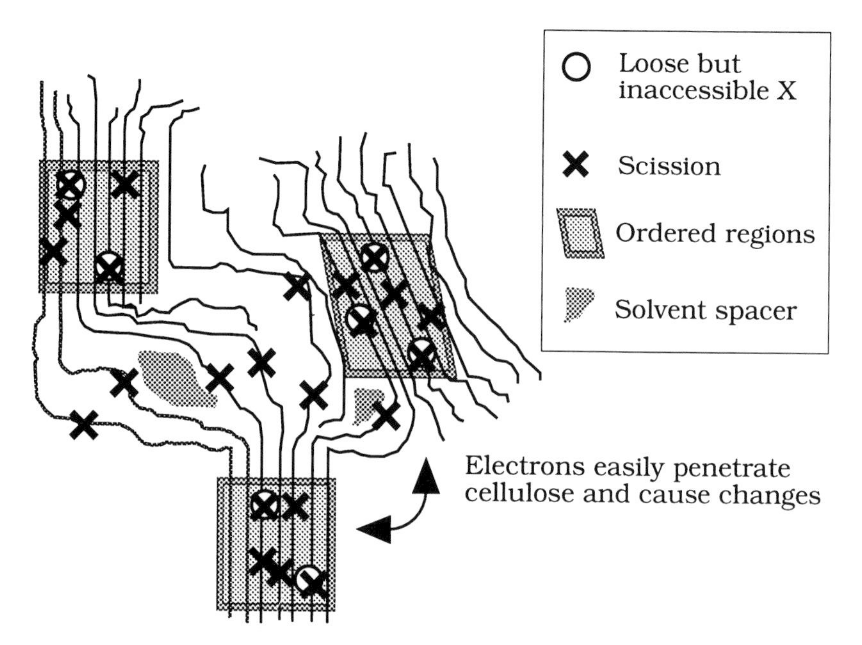

8.4 Supermolecular changes induced by electron processing (lateral view).

completely and the concentration of carbon disulphide reduced from the norm of 30–36 down to 18–26%. The unreacted cellulose which in the traditional process causes filtration problems is also removed. Figure 8.5 compares the effects of the electron treatment on each of the steps in the traditional viscose process. The benefits, therefore, are:

1 Significant chemical saving on carbon disulphide, sodium hydroxide and sulphuric acid.
2 Better product due to homogeneous xanthation.
3 Better process control.
4 Great environmental advantages.

In short, electron processing enhances cellulose reactivity, increases filterability, improves viscosity and allows the utilisation of a greatly reduced concentration of chemical reagents. These chemical savings amount to US$6 million per plant (60,000 tonnes per year) for carbon disulphide and US$0.5 million for alkali per year. Even greater, perhaps, in the current climate towards chemical pulping and viscose plants are the reduction of 40–50% in sulphide emissions, which also greatly reduce the clean-up costs. There is elimination also of the traditional ageing step. Undoubtedly, there are important advantages to be gained from introducing this new technology.

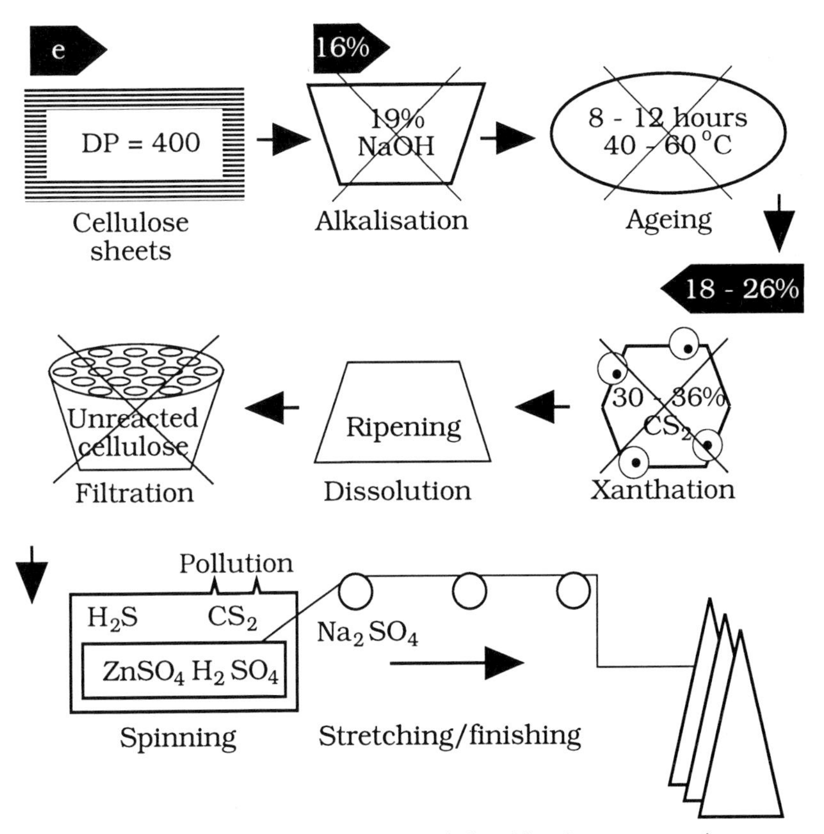

8.5 Changes to the viscose process induced by electron processing.

Note: cross through step indicates either elimination of the step or reduction in concentrations used after electron processing.

8.3.2 A total chlorine-free pulping process

According to public opinion, the production of pulp and viscose fibers is associated with heavy pollution of air and water by emissions of sulphur dioxide, hydrogen sulphide, carbon disulphide, chemical and biological oxygen demands (COD and BOD), and halogenated organic compounds (AOX). This aspect has proved the Achilles' heel of the man-made cellulosic fiber industry compared with synthetic fibers and the natural fibers. On a raw material basis, the variability in physical properties using fiber engineering and the problems associated with disposal, the new cellulosics are now approaching the stage where they can hold their own with their competitors. The pollution problem is, therefore, fundamental for their future well-being in the market-place. Reduction of emissions, the avoidance of hazardous chemicals, the economical use of limited resources and the development of new environmentally sound technologies present an important challenge in order to improve the position of viscose within the inter-fiber competition.

The Lenzing Company of Austria has shown the way forward. No other company could possibly have faced such stringent regulatory requirements as were imposed upon it in Austria. In response, Lenzing developed a unique strategy to meet the very strict requirements of the Austrian authorities. A waste-water project and a clean air programme were started in the early 1980s to reduce pollution drastically. But in the new circumstances that was not good enough. New technologies, unavailable at the time, had to be invented. Lenzing invested US$300 million and had to reduce its workforce by a third from 3,800 to 3,000 to retain its competitiveness. Now new key technologies are in place: the vapour condensate extraction, the medium-consistency ozone bleaching, and the thermal monosulphite splitting process all had to be developed *ab initio*. As a result Lenzing is now able to produce viscose fibers, which are totally chlorine-free and create a minimum of pollution during the production process. Such fibers are now considered to be the optimum raw material for medical and hygienic products.

The transformation has been remarkable. In 1982 Lenzing's waste-water load from pulp, paper and viscose production still equalled a population equivalent of more than 1 million, which was an acceptable value at the time. Within a decade it has been possible to reduce the waste-water load to 3,000 population equivalences, which is less than 1% of the initial value. Recovery of all materials used has been the key. For example, investment in a new recovery boiler with highly efficient flue gas desulphurisation reduced sulphur dioxide emissions from the pulp mill, energy and sulphuric acid production to about 20% of the 1985 value. The improvement was even more dramatic with regard to the odour-intensive component hydrogen sulphide. After starting up a new Sulfosorbon plant for carbon disulphide and sulphur recovery from lean gases and utilisation of hydrogen sulphide-rich strong gases for sulphuric acid production, hydrogen sulphide emission dropped to 2.5% of the 1985 value.

Energy savings too have been dramatic. Owing to the interlinked power economy of the pulp and the viscose factory and the thermal utilisation of residual substances from the process, such as bark, thick liquor, biological sludge, it is almost possible to eliminate the need for the use of fossil fuels.

The elements in this clean production process are:

1 Closed loop operation of spin bath and stretch bath.
2 High-yield recovery of sodium sulphate.
3 Steeping-lye purification by dialysis.
4 Incineration of waste lye with soda recovery.
5 Carbon disulphide and sulphur recovery in sorbon plants.

6 Production of sulphuric acid from hydrogen sulphide-rich gases.
7 Chlorine-free fiber bleaching.
8 Biodegradable additives.
9 Removal of zinc from effluents by precipitation.
10 Biological waste-water treatment.

The overall consequence is a chlorine-free viscose fiber produced with environmentally sound technology. It meets the requirements of hygienic fiber consumers, and it is part of an ecologically closed life cycle based on the natural, replenishable raw material wood. This praiseworthy development points the way forward for the industry as a whole. The problems remain enormous in developing countries, particularly where the sulphate pulping Kraft process remains the most viable option with chlorine dioxide used to bleach the pulp. Increased environmental awareness will surely demand that all countries approach pollution control in global partnership.

8.4 Life-cycle assessment

The growing public and industrial interest in environmental issues has led to the development of different methods for the assessment of the environmental impacts from materials, products, processes and activities. A widely used method is the so-called life-cycle assessment (LCA). The purpose of an LCA is to quantify the environment burden from cradle to grave for a production system, including extraction of raw materials, processes, transports, use and final waste disposal, as illustrated in Fig. 8.6.

The Akzo Nobel AB Company has carried out an assessment of the environmental friendliness of the man-made cellulosics in comparison with the polyolefines. The amount of detail required for this analysis is quite extraordinary. It must be emphasised that the results are only relevant to the particular production system and the conditions pertaining to that country and site location. First, a flow chart of the life-cycle system for each product must be drawn up, starting with the initial natural resources.

It is tempting to ask whether the cellulosic fibers are more environmentally friendly than the polyolefines. The Akzo Nobel AB study carried out this comparison, starting from the crude oil for polypropylene fibers and in the forests for viscose. From these natural resources, raw materials are extracted and used in various formulations to produce consumer products, some of which can be refused or recycled before being converted into waste. All stages in this life cycle contribute to the total environmental burden and must be taken into account. For an LCA, a standard structure laid down by SETAC, the Society of Environmental Toxicology and Chemistry, must be adhered to.

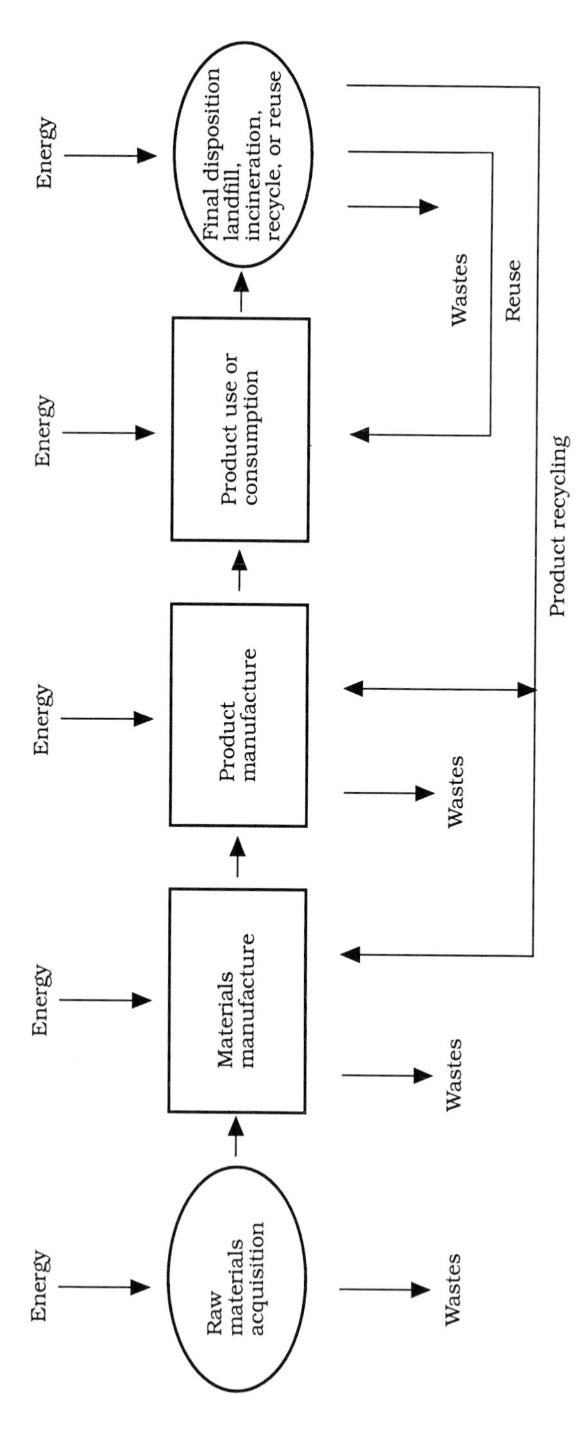

8.6 General materials flow for "cradle-to-grave" analysis of a product distribution system.

The individual components that must be quantitatively considered for the production systems of viscose and polypropylene are shown in Fig. 8.7 and 8.8.

Only the preliminary results are available, since the study has not yet been completed. It is not possible to answer the basic question unequivocally and decide which of the two products is most environmentally friendly in all aspects. However, the initial results are indicative of the different environmental impacts of the two products, and then, it must be emphasised, only for site-specific conditions in Sweden for the viscose data and an average value for European conditions adopted for polypropylene.

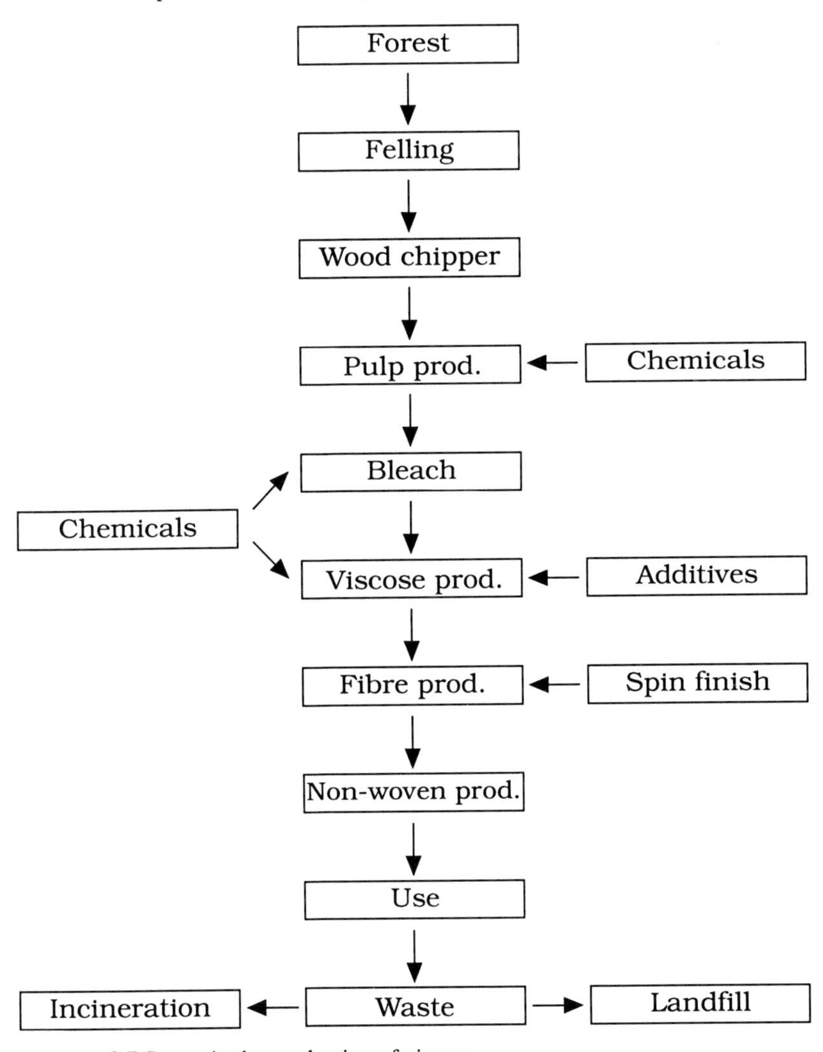

8.7 Stages in the production of viscose.

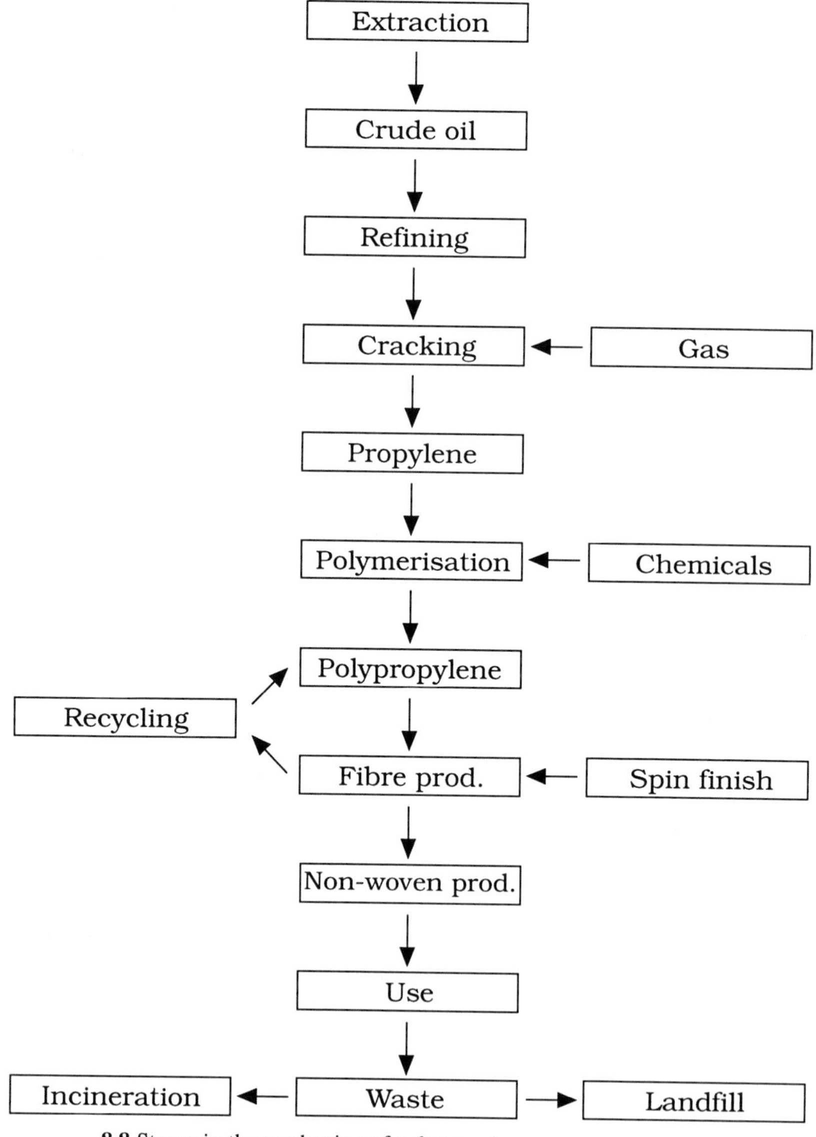

8.8 Stages in the production of polypropylene.

The total energy consumption is comparable for the two systems, including inherent energy of feedstock materials, but for viscose fiber almost half the total energy is based on renewable materials. For carbon dioxide which originates from renewable resources there is no net effect on the *global warming potential*, which is the situation that pertains for viscose, but not for polypropylene. Viscose, on the other hand has a greater impact on acidification owing to the emission of hydrogen sulphide, carbon disulphide and the sulphur and nitrogen oxides. However, greater amounts of nitrogen oxides and hydrocarbons are generated by polypropylene production. Undoubtedly, polypropylene production has a greater impact on photochemical oxidation. On the basis of the Swedish management it is more advantageous to incinerate both polypropylene and viscose with energy recovery than to deposit in landfills. However, since polypropylene has a higher heat of combustion than viscose, the energy recovery is higher. Overall, however, when considering the environmental impact based on global warming potential that might be expected over the next 20 and 100 years, the *photochemical ozone creation potential* caused by outlets of hydrocarbons, eutrophication brought about by emissions of nitrogen oxides and *acidification* brought about by sulphide, nitrogen and sulphur oxides emissions, it is the polypropylene that imposes the greatest environmental impact.

This type of analysis will inevitably be required for most products in the future as concern for the environment grows. Here the natural and modified cellulosics have a built-in advantage which they must energetically exploit commercially. The information can be used for fulfilling the need for eco-labelling criteria and as a basis for strategic planning for future product developments.

8.5 Cellulose: the renewable resource

It is important to recognise that the available resources of cellulose from wood represent a major supplement to oil-based polymers, and could well provide a means of tackling the inevitable critical problem of clothing the world in the future when oil is becoming depleted. There is now very active and successful research into developing fast-growing trees. From these, new fiber materials could be produced. The features of cellulose which can support such a development can be summarised:

1 Rigid segments that give good fiber-forming ability and high Young's modulus.
2 Various functionalities can be introduced by the controlled rearrangement of hydrogen bonds.
3 Moderate wettability and humidity-retaining ability provide comfortable clothing for the human body.
4 Superior biodegradability is a built-in factor when considering environmental protection.
5 The cellulose structure lends itself excellently to chemical derivatisation and so allow the ability to introduce special qualities.

Table 8.2 illustrates the already extensive range of applications, which will surely multiply in the years ahead.

Table 8.2. Cellulose and its derivatives

Type	Material	Application
Celluloses	Cotton, linen	Fiber, non-woven fabrics
	Powdered cellulose	Filter
	Micocrystal cellulose	Medical, chromatography
	Microfibril cellulose	Foods, cosmetics
Regenerated cellulose	Viscose	Fiber, tyre yarn
	Benberg	Fiber, dialyser
	Tencel	Fiber
	Spherical cellulose	Chromatography, cosmetics
	Cellophane	Food-wrapping film
	Sponge, non-woven fabrics	Domestic goods
Esters	Cellulose acetate	Fiber, film
	Cellulose nitrate	Paint, gunpowder
	Cellulose acetate phthalate	Medical (masking)
	Cellulose acetate buthylate	Plastics
Ethers	Methylcellulose	Cement mixture, sizing agent
	Ethylcellulose	Lacquer, paint
	Hydroxyethylcellulose (HEC)	Paint, latex
	Hydroxypropylcellulose (HPC)	Medical (adhesive), cosmetic
	Carboxymethylcellulose (CMC)	Foods, binder, adhesive
Etheresters	Various	Medical (masking)

9 Fibers for the next millennium

To look forward, it is necessary to look back and summarise the fiber developments that have been described in previous chapters. These are illustrated in Fig. 7.5. The key objectives have been to improve the performance, function and productivity of fibers. To achieve these aims, innovative technology was required. For the development of fibers with strength, super-high tenacity, and super-high modulus, new technologies needed to be developed. To reach the ultimate in fineness, a radically new approach was required. Since these will need to be the springboard for new innovations, the principles used should be identified.* On this foundation, the leap into the next millennium can be taken (Fig. 9.1).

9.1 High-tenacity and high-modulus fibers

Four new fiber-making technologies have produced the "super-fibers", which need to have a tenacity of at least 20 g/denier, to qualify for this category. First is gel-spinning, super-drawing fiber-making technology that, after forming gel-like fibers by wet-spinning a super-high molecular weight polyethylene solution, draws out the fibers at a very high draw-ratio (Fig. 9.2a). Fibers produced by this process have greater tenacity and a higher modulus than any other organic fibers currently commercially available. Because of their low melting point, however, these fibers have only limited uses.

The second technology is a liquid crystal spinning process (Fig. 9.2b), which spins a liquid crystal solution of rigid polymers in a semi-dry and semi-wet state and then produces highly oriented crystallisation of rigid polymers by a spinning draft. The process uses a high concentration of sulphuric acid, as a solvent. Among typical examples of products using this process are the

* We acknowledge the material and illustrations from Teijin Ltd which have been incorporated into this chapter.

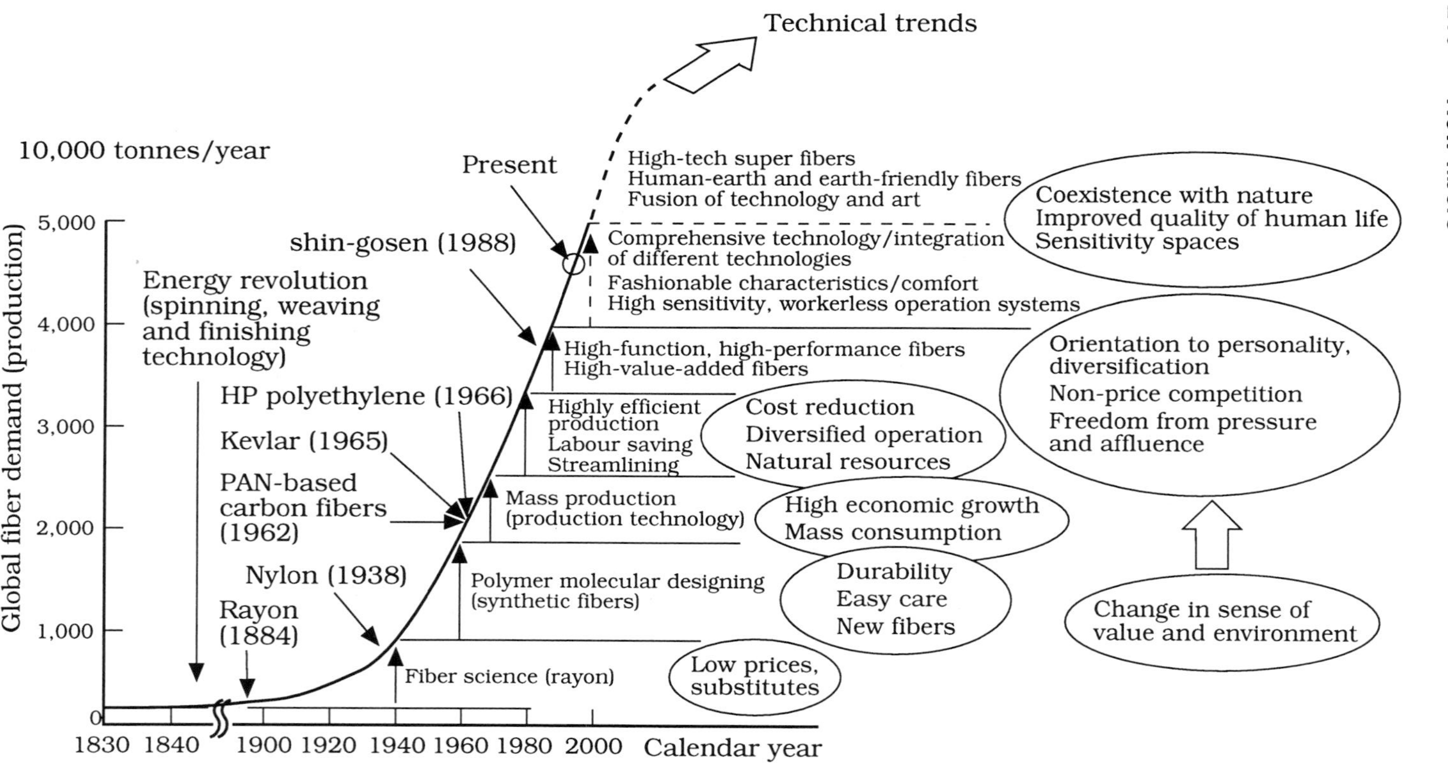

9.1 Factors of fiber technology and trends of fiber demand (production).

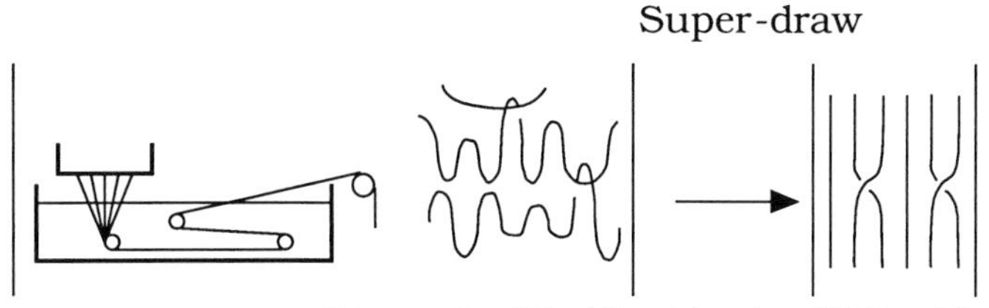

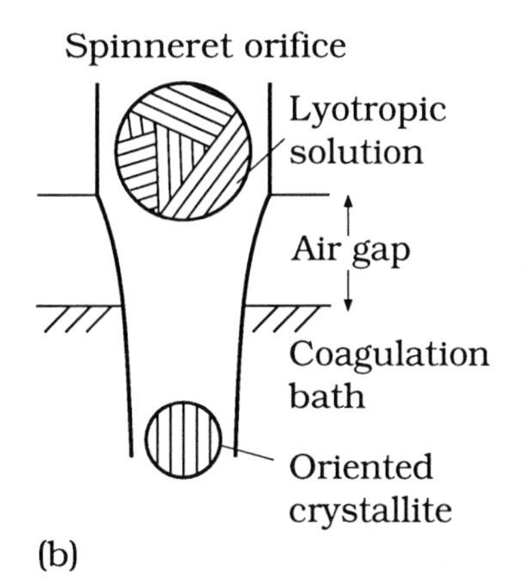

9.2 (a) Advanced technology for molecular orientation (gel-spinning and super-drawing). (b) Spinning of anisotropic polymer solutions.

para-aramid fibers such as polyparaphenylene terephthalic amide (PPTA). Aramid fibers have high tenacity and great heat resistance.

Another new product using this process is poly-*para*-phenylene bisoxazole fiber (PBO fiber), which has ultimate values in the modulus and tenacity. This can be regarded as one of the products for the 21st century. Stanford Research Institute acquired the basic patent, and Dow Chemical (USA) purchased the patent, and have enlisted the Toyobo Company (Japan), who now have developed a pilot plant for its production. It is a wonder fiber, stronger than steel, superior to carbon fibers, with twice the strength of Kevlar. A single fiber, a mere 1 mm in diameter is strong enough to lift 400 kg (the weight of a cow). Commercial production will start in 1998. PBO is polymerised from diaminoresocinol dichloride and terephthalic acid in polyphosphoric acid. PBO's second outstanding feature is its high Young's modulus, exceeding twice that of Kevlar. Most materials of high strength and Young's modulus,

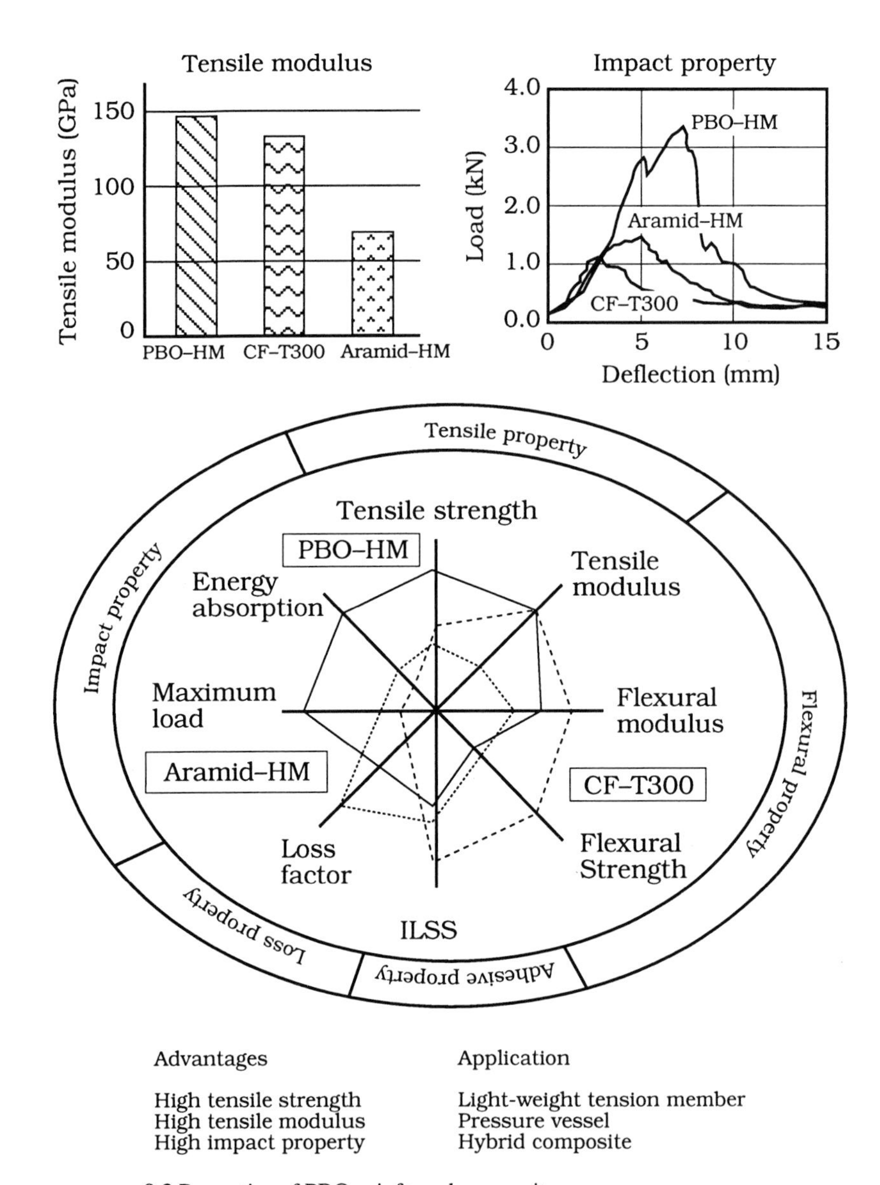

9.3 Properties of PBO-reinforced composite.

like carbon fiber, are quite brittle, but PBO is strong, yet flexible. Its third important characteristic is its flame resistance. Flame resistance is measured by the limiting oxygen index, which is 56 for PBO, markedly greater than polybenzimidazole, the former record breaker at 42. Applications, well into the 21st century, could be as a heat-resistant cushion for aluminium and glass manufacturing processes, tension members for optical cable and cord,

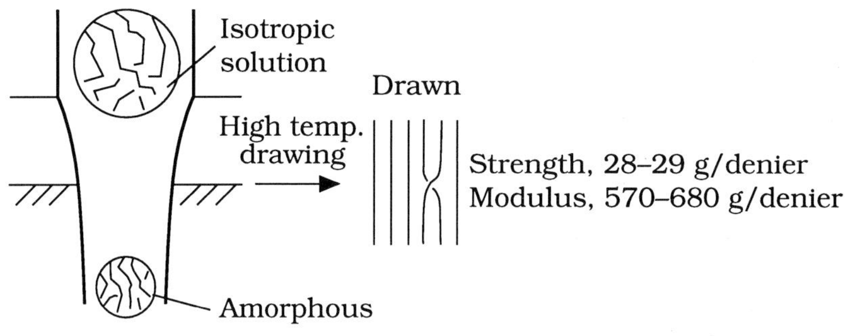

9.4 Spinning of isotropic polymer solutions.

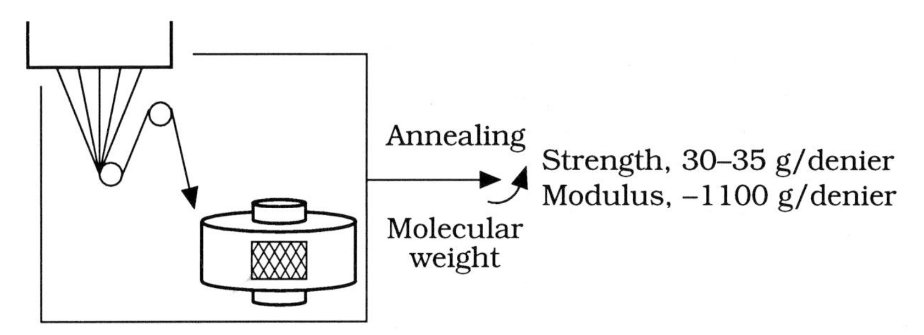

Results
Modulus has almost reached theoretical value (ca. 80%)
Strength is 1/10 of theoretical value
A wide range of applications is expected

9.5 Melt spinning of thermotropic liquid crystal polymer.

composite cables for bridges, turbine engine fragment containing, etc. It is truly a wonder fiber. The properties are summarised in Fig. 9.3.

Diaminoresorcinal dichloride (HO, OH, H_2N, NH_2) · 2 HCl + HOOC–(Terephthalic acid)–COOH → [PBO]

The third technology reforms the rigid *para*-aramid molecular structure and spins in semi-dry and semi-wet systems by dissolving in an organic solvent. Although the feed stock is in an amorphous state at the spinning stage, this new spinning technology highly orients molecules by hot-drawing at high temperature (Fig. 9.4). This technology uses an organic solvent instead of high-concentration sulphuric acid. The product has a greater tenacity than aramid fibers made by the liquid crystal spinning process, but its modulus is limited.

The fourth process gives super-high-tenacity to the fiber by melt-spin of semi-rigid polymers through heat treatment (Fig. 9.5). This technology, specifically for aromatic polyesters, does not use a solvent.

These new technologies have respective special features, and physically, some of them have provided a means of achieving a modulus closer to the theoretical value (i.e. *ca.* 70–80%), but in terms of tenacity, they have only attained one-tenth of the theoretical value. To achieve the ultimate in strength, more work is needed, but from a practical point of view, it may be argued that super-high-tenacity and high-modulus yarn-making technologies have already been established.

9.2 Microdenier (ultra-fine) fibers and biomimetics

Manufacturing processes can now make fibers of 0.0001 denier. Typical ultra-fine yarn-making technologies available can now be summarised. In this respect, learning from nature (biomimetics) has assisted greatly in planning fiber design. Natural fibers have built into them structures that enable them to perform specific functions within human and animal bodies. By replicating these technologically, high-performance, superior functions and high aesthetics can be achieved.

There are four methods to produce ultra-fine fibers as shown in Fig. 9.6. Direct spinning is simply an extension of conventional spinning, and has a technical limit with regard to the denier. The islands-in-the-sea type produces ultra-fine fiber of 0.0001 denier. The separate type allows ultra-fine fiber to be produced with a sharp edged cross-section and the multi-layer produces flat fibers of mixed denier. The elegant fibers that can be produced by these methods are shown in Fig. 9.7, which is a peach-skin type fabric, composed of ultra-fine fiber and trilobal cross-section fiber in filling and warp respectively. Figure 9.8 shows the new spun-type yarn composed of fine fiber as a sheath and thick conjugate fiber as a core.

To produce a synthetic with all the qualities of silk has proved a continuing challenge to the fiber scientist and technologist. It has now been achieved by a variety of technological innovations successively introduced. The silk features that have been replicated in synthetic fibers and the methods used to achieve them are:

1 Lustre – triangular cross-section.
2 Drapability – lowering contact pressure between yarns by alkali weight reduction.
3 Soft feel – ultra-fine fiber.
4 Bulkiness – Mixed weaving, and combining fibers with self-extensionable yarn.

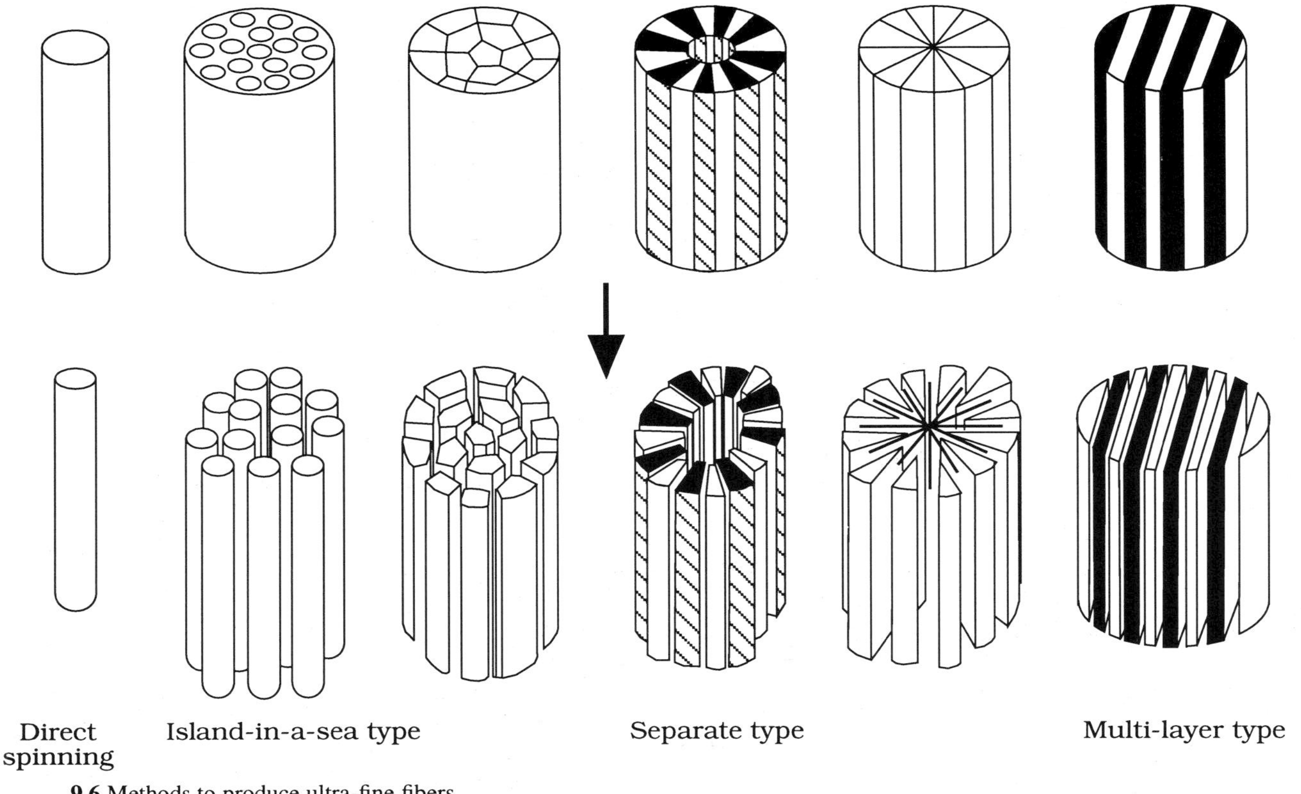

9.6 Methods to produce ultra-fine fibers.

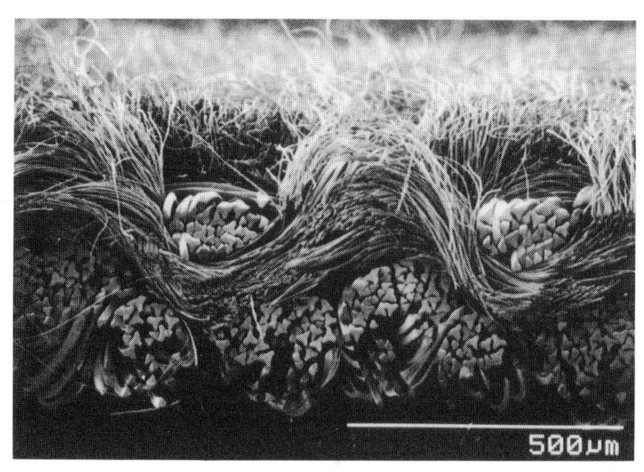

9.7 Fabric from various ultra-fine fibers.

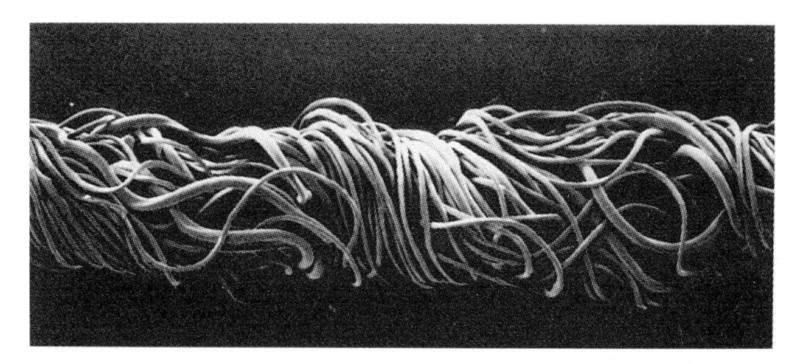

9.8 Yarn from fine fiber sheath and a conjugate fiber core.

5 Scrooping sound – introducing irregular shape and microgrooves.
6 Natural appearance – Combining various deniers and modifying and touching shapes in cross-section and/or combining filaments and staples.

Further improvement can be expected in the future by controlling fibrillar structure and moisture absorbency through amino acid compositional changes and fibrillar structure.

It is the bulkiness of wool that has been emulated, achieved naturally through its macrostructure which comprises crimps and staple fibers, resulting from bilateral structures consisting of *ortho*-cortex and *para*-cortex. To copy the crimps in wool, polyester fibers were similarly crimped using false-twisting technology, making the most of their thermoplasticity. A process was developed to give these polyester fibers a spun-yarn touch similar to wool by forming micro-loops through air texturising or by making the surface fluffy by nap-raising. The most difficult task was to achieve the opposing characteristics of wool, namely stiffness and resilience, but remaining soft to the touch. This was eventually achieved by simultaneously false-twisting

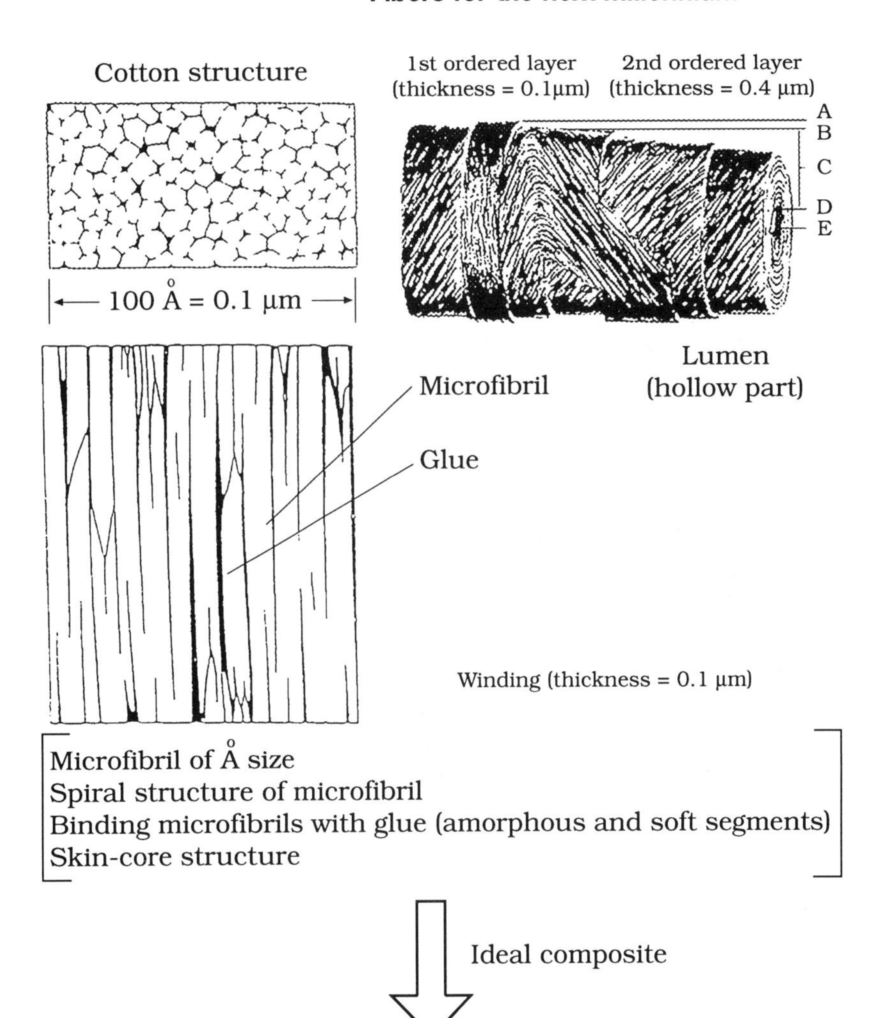

9.9 Future innovations: ideal composite structure – cotton.

filaments with different elongation. Wool also has excellent shape-recovery properties arising from the cuticle structure on its surface and the complex islands-in-the-sea and skin–core structures specific to wool. There are other characteristics, too, that have not yet been fully replicated, such as combined water-repellancy and moisture-absorption.

Cotton remains the most widely used of all fibres, and has posed complex problems in its improvement and duplication synthetically. The water-absorbing properties, which make cotton so desirable in warm climates, arise from micro-holes and lumens in its structure (Fig. 9.9). The challenge in the future is to replicate this structure, which is an ideal composite, as illustrated in Fig. 9.9. Some of these properties can be introduced into synthetics by emulation:

1 Sweat/liquid absorbency – modified cross section, and introducing porous/hollow fiber.
2 Warmth retention – high degree content of hollow filament.

However, synthetics have not been able to replace cotton and the new man-made cellulosics (Chapter 8) appear now to be the best method of obtaining the combined properties of cotton and the synthetics, rather than trying to modify the synthetics to copy cotton. There have been other properties introduced into synthetics not provided by the natural fibers, but these can be combined with those features that have been described. They include combined moisture permeability and waterproofing, electrical conductivity, antibacterial and odour-preventive qualities, fragrance and wood-aroma, stretchability, ultraviolet radiation resistance, intra-hospital infection capability and ultra-light weight.

9.3 The next stage: technological improvements

Certain technological innovations can be envisaged, although they are not yet fully achieved. The reduction in cost and improvement in productivity are two tasks that must be successfully addressed. Already much has been accomplished. In polymerisation, for example, continuous processes have expanded production from 100 tonnes in 1986 to 300 tonnes in 1996. Speed of filament spinning has increased from 4,000 m/min in 1986 to 7,000 in 1996. In the staple fiber sector, the main progress has been in the expansion of capacity, rather than in spinning speed, and capacity has been raised from 70 tonnes per day in 1986 to 130 in 1996. Meanwhile, the texturing speed for draw-textured yarn has increased from 700 m/min to 1000.

Rationalisation of the manufacturing process is a vital part of reducing production costs. Previously, polymerisation, spinning, drawing and texturing were performed by four independent processes. Now, continuous spinning and direct drawing technology consolidate the spinning and drawing processes. Production of textured yarn is now rationalised, with either the polymerisation and spinning stages directly connected or with unified drawing and texturing processes. Meanwhile a large-capacity spinning process, directly connected with continuous polymerisation, has been adopted for staple fiber production, resulting in significant cost reductions. The manufacturing process for filaments has already been developed into a continuous polymerisation, direct spinning and direct-drawing technology that has all three stages linked. Now, also, an integrated process comprising polymerisation, drawing and texturing has been developed in nylon carpet production.

Simplification of manufacturing processes has been achieved, particularly in high-speed spinning and direct fabrication of sheet fabrics. Spun bond and melt blow technologies are now available which directly gather spun filaments or fibers on the net and directly fabricate them into non-woven fabrics. These processes will surely be extended further in the years ahead.

Further possible improvements in the manufacturing technologies are illustrated in Fig. 9.10, which push forward the improvements already evident. Automation and a stable trouble-free technology, with flexibility, will be required.

9.4 The next century: respect for people's quality of life and harmony with nature

In the past, priority was given to the objectives of manufacturers, including highly efficient mass production, but in the future, more attention must be given to consumers' interest and the quality of people's life. Harmony with nature should also be taken into consideration. A problem to be solved is how to incorporate those factors into business. Research emphasis will inevitably be shifted from hardware to software. Figure 9.11 illustrates the concept of the dimensional factors for the next generation of fibers. Essential to progress is a partnership between fiber science and technology, industry and education. Figure 9.12 illustrates the fundamental directions that fiber science must encompass in the next generation if the new concepts are to be introduced into fiber design and manufacture.

9.4.1 Dimensions and structural control of fibers

A fiber can presently be defined as the unit composing yarn, fabric, etc. which has a sufficient length relative to thickness and is thin and easy to bend. This agrees with the macroscopic definition of fibers that are used, not only for clothing, but also for medical and industrial purposes. For example "dietary fiber" consists of a staple so minute that it cannot be seen with the naked eye; artificial kidneys use hollow fibers. Super-fibers stronger that steel wire have been developed and are used in the fields of sport, recreation and aerospace. The word "fiber" encompasses not only a mass of threads but nerve and

Key points:
- Super high-performance technologies *(basis for labour saving)*
- Flexible manufacturing system
- Efficient value-added technologies

1 Super high-performance technologies

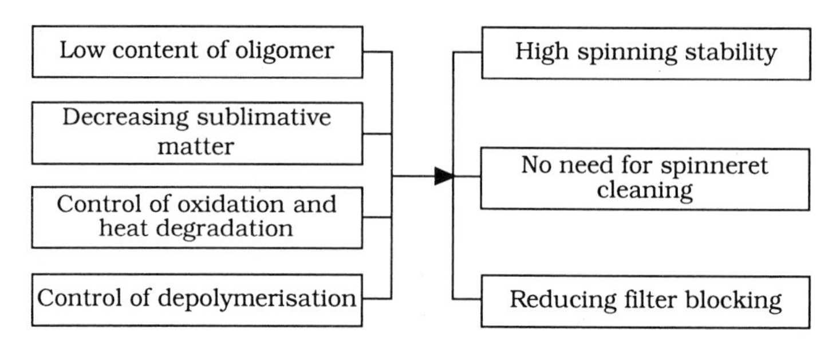

2 Flexible manufacturing system

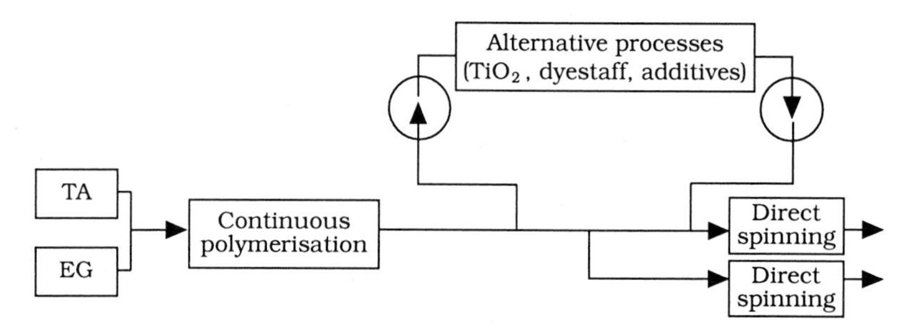

3 Efficient value-added technologies (polymer co-spinning system)

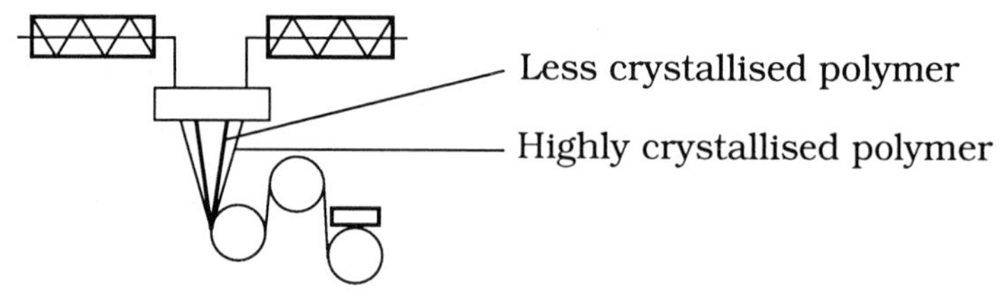

Mixed fiber by one shot process

9.10 Future improvements of production technologies.

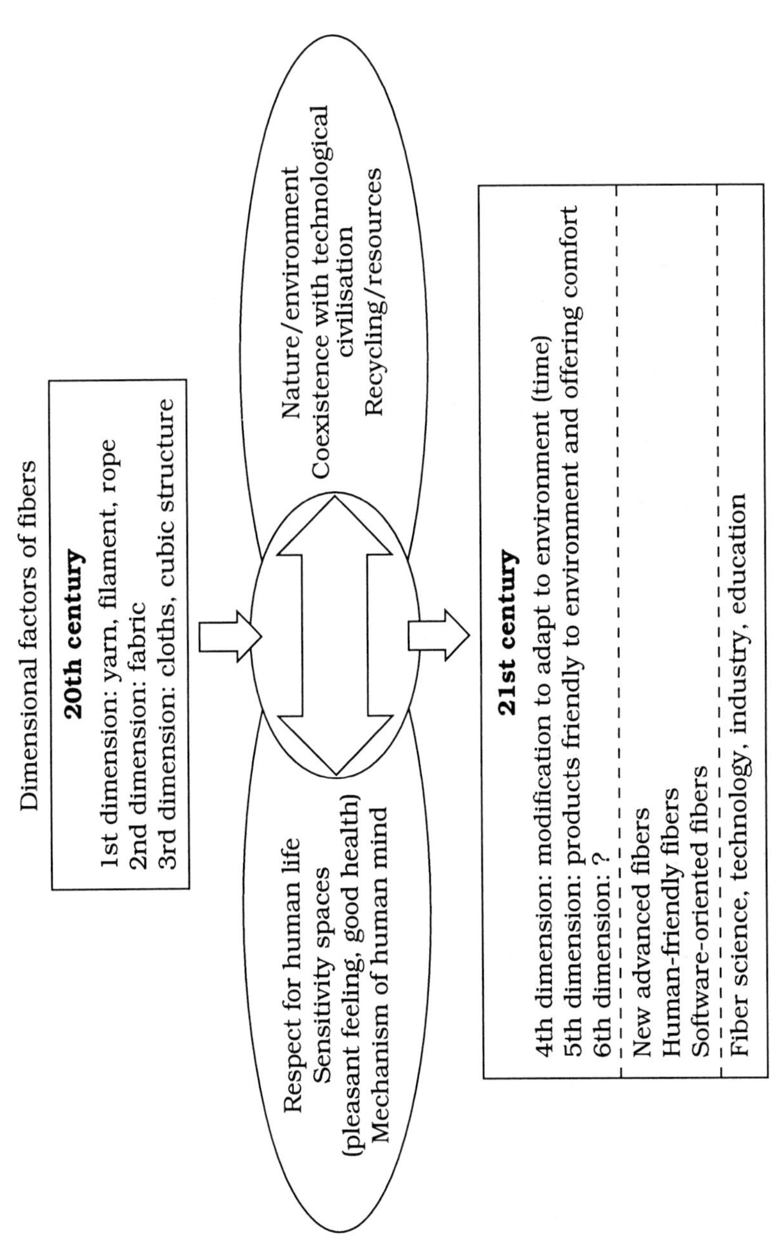

9.11 Concept of dimensional factors of next-generation fibers.

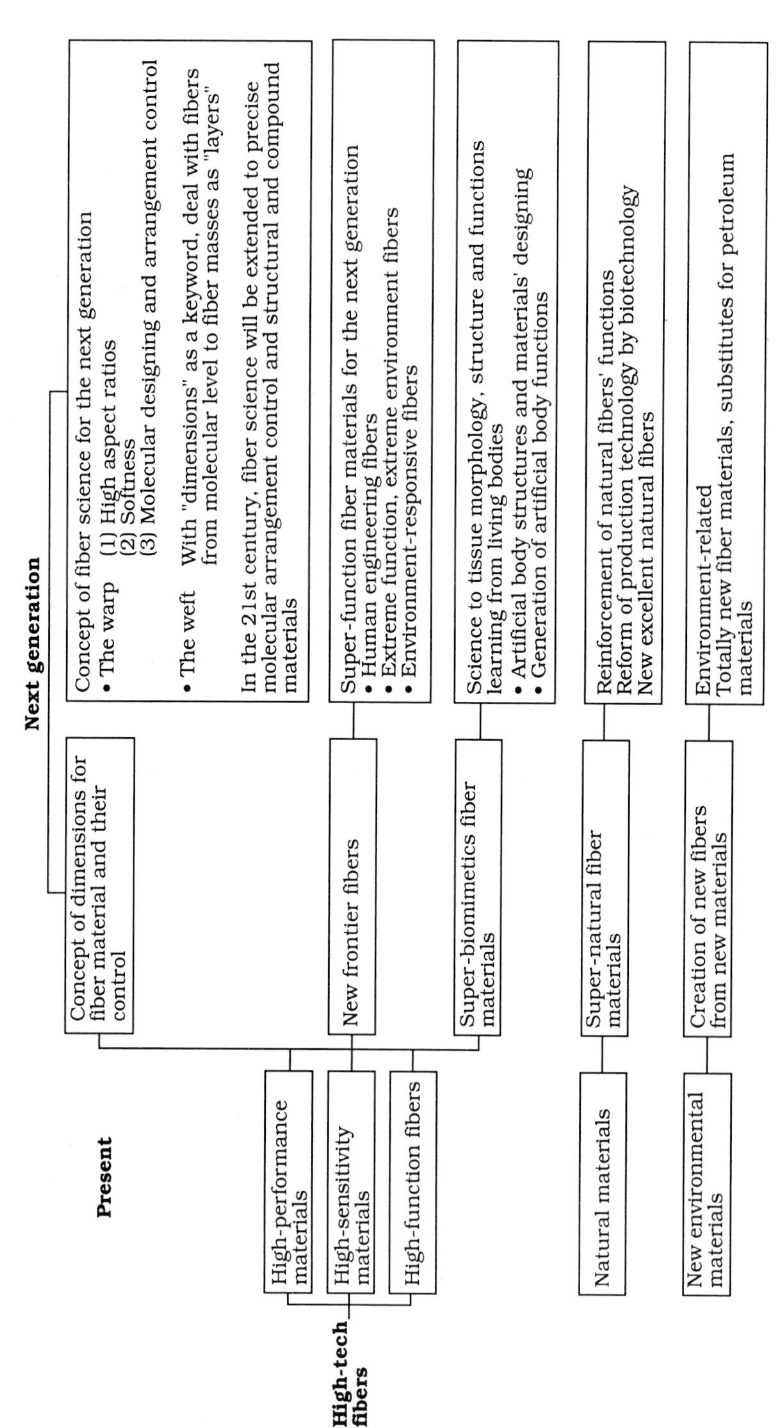

9.12 Fundamental directions of fiber science for the next generation.

muscle fibers also. Thus the element of "size" from the term "fiber" can be removed and enable redefinition of the term as "a thin and long substance with strength and elongation of a certain level or greater". Because it is by definition thin, the fiber is able to bend (degree of freedom). To characterise the thin and long character of fibers numerically the ratio (L/D) of their length (L) to their diameter (D) is useful; this is known as the aspect ratio. The fiber can be defined as a material that has an aspect ratio of 100 or more.

The diameter of existing fibers ranges from a few to scores of μm. If we apply the definition of fiber with aspect ratios down to molecular levels, we can consider the possibility of fibers with a diameter of the order of an angstrom. Such fibers are represented by linear polymers, such as cellulose and polyethylene which are referred to as molecular fibers. DNA is similarly referred to as a nano-fiber. Thus, the term fiber can adequately describe a range of thicknesses ranging from molecular polymers (1–10 Å) to the thinnest fibers that can be spun (1 μm) to twine (1 mm) and hawsers for ships (100 mm).

Within this broad fiber definition, three targets must be set: (1) exploitation of the morphological characteristics of fibers, (2) scientific and technological pursuit of softness; and (3) extend the performance and function of fibers by molecular design to control the warp and dimensional structure.

9.4.2 Next-generation fibers and dimension

9.4.2.1 Fibers as first-dimensional materials

Because fibers have the morphological characteristics of being thin and long, they are referred to as the first-dimensional materials. To exploit the characteristics of these first-dimensional materials thoroughly is one of the directions fiber science must take in the future. For example, an extremely fine filament has been experimentally produced, which can reach the moon (384,400 km) with a total weight of only 4.16 g. Other new filaments can envelop the Earth with a weight of only 1 g. Researchers, however, have not yet controlled the boundary between the macroscopic fiber structure and chain-like high polymer molecules composing it, nor have they established the technology for isolating molecular chains and dealing with them as independent fibers. We still do not know what characteristics will emerge in the property of the fiber materials when the fibers become much finer and increase their surface and interface.

9.4.2.2 Fibers as high-dimensional materials

The techniques for providing fibers with a higher-dimension aspect has been exploited for the production of synthetic fibers, where cross-section is modified into a non-circular shape or is made of multiple components. For example, non-see-through white bathing suits are made of the new fabric produced by combining several different materials and modifying the shape of fiber cross-section. By applying the idea of higher dimensions to the structure, like the scales of wool, it might be possible to create a synthetic fiber with a different frictional coefficient according to the direction of the force. The higher dimension of fibers can be realised by modifying the molecular arrangement in a specific manner. If the molecular arrangement can be modified specifically according to the location of molecules in the cross-section, the mechanical property of the fiber can be designed at its molecular level.

In future, new fiber materials could be produced where the cross-section has a slanting or non-continuous layer structure by arranging molecules specifically in a radial direction of the fiber cross-section or by combining structure-controlled fiber materials.

9.4.2.3 Super-dimensional fiber science

It is also important to expand the potential of fiber materials by introducing the concept of time and space dimensions. The ultimate goal is to provide fibers with intelligence that could introduce sensor-actuator function to fibers. To do this, a system with a particular purpose must be built into the fiber materials. Here we need to integrate a broad range of knowledge and technology, covering the research areas of molecular design, molecular composites and high-order structural control of fiber materials.

Since fibers are regarded as the materials of sensibility, a new type of technology should be developed to exploit the full capacity of fibers. Here human *Kansei* (*Aesthetic*), corresponding to aesthetic engineering must be evaluated quantitatively, and then correlated with physical quantities specifying fiber characteristics. Conventional physical quantities will not be sufficient to specify the fiber characteristics fully, and a new concept such as fluctuation, fractal or chaos will be required to extend the range of observable physical quantities. In other words, the technological strategy should be shifted from hardware-oriented to software-oriented. The software-oriented technology may be able to create novel fibers of high sensibility which we cannot fully conceive at present.

9.4.3 Systemisation of software fiber science

Today's consumers demand more of what will enrich their life. They desire new standards in areas such as health, amenity and education. In the next

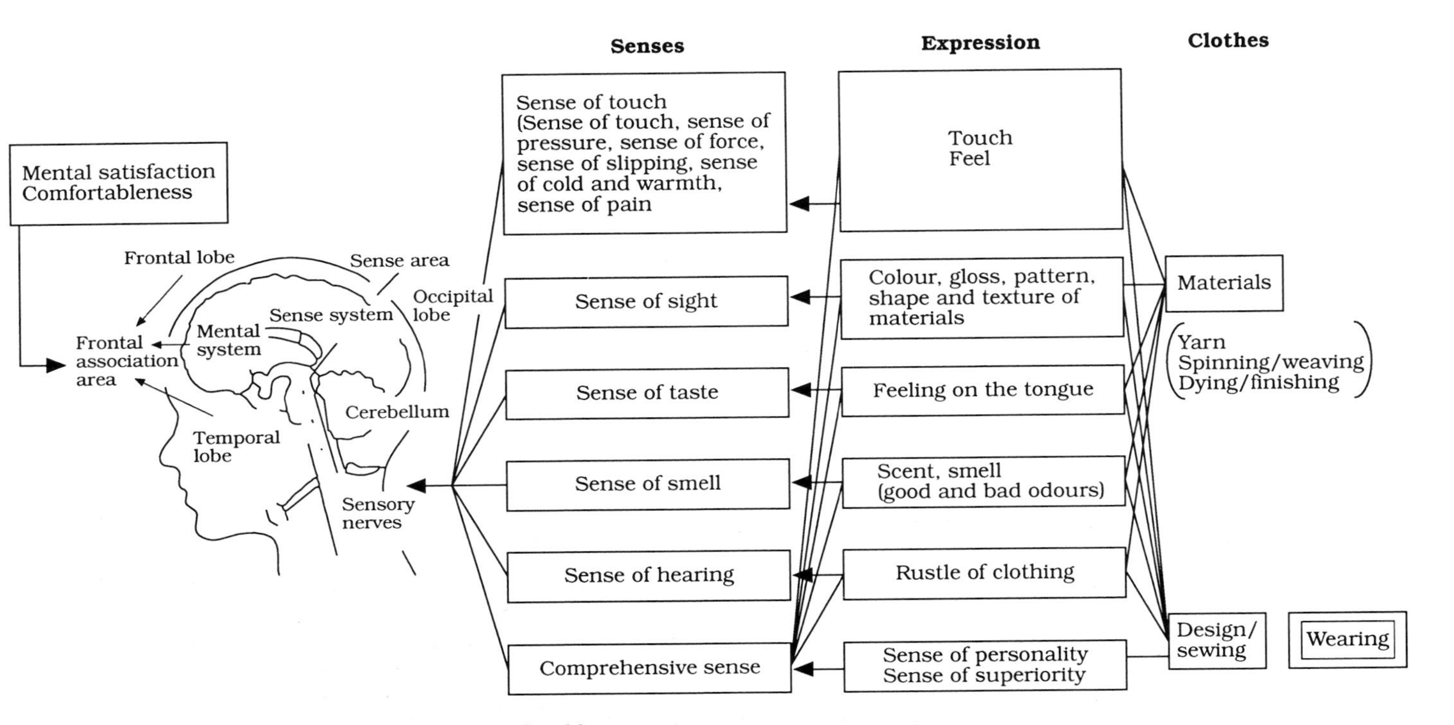

9.13 Mechanism by which a suit is felt comfortable to wear.

generation, such unprecedented new aspirations may determine the quality of fiber materials, so that a new system needs to be developed urgently to evaluate human sensibility.

"*Kansei* (Aesthetic) engineering" has now become a full-fledged field of science in Japan in order to deal with human mental activities such as the appreciation of beauty. People are becoming more interested in the pursuit of sensibility. Technology and art became two independent subjects with the development of modern industries, but now they are expected to amalgamate to create new commercial values.

The scientific study of the mechanism of the human mind includes mental activities for appreciating aesthetic beauty. We interact with the environment through sensory organs, which input outside information, conveyed by the nervous system to the brain to be processed and fed back. We feel comfortable when such processed information is recognised as being agreeable, but there seems to be no definite criterion which fluctuates according to the mental state. So it becomes important to develop a system for value judgement through the analogous cycle:

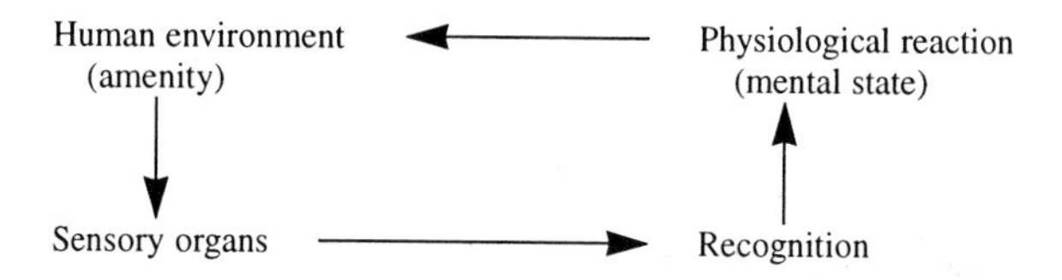

This cycle may be a model of the mechanism by which we feel when a suit is comfortable to wear. If we understand the mechanism, we can evaluate the comfort of clothes and other environmental factors quantitatively. Figure 9.13 illustrates this mechanism.

9.5 New frontier fibers (super-function fiber materials, etc.)

The term "new frontier fibers" implies opening up a new area of advanced high-tech fibers. These next generation fibers should be those products whose functions are highly designed and controlled by the fusion of fiber science and other fields and should have tertiary functions, that is, super-functions. Examples include intelligent fiber materials that can control their own functions according to the environment and "super-function fiber materials," such as optical fibers, in which functionality is incorporated into the fiber. Figure 9.14 shows the development steps necessary to produce such functional fibers for the next generation.

In a society where human needs have priority, attention must be paid to

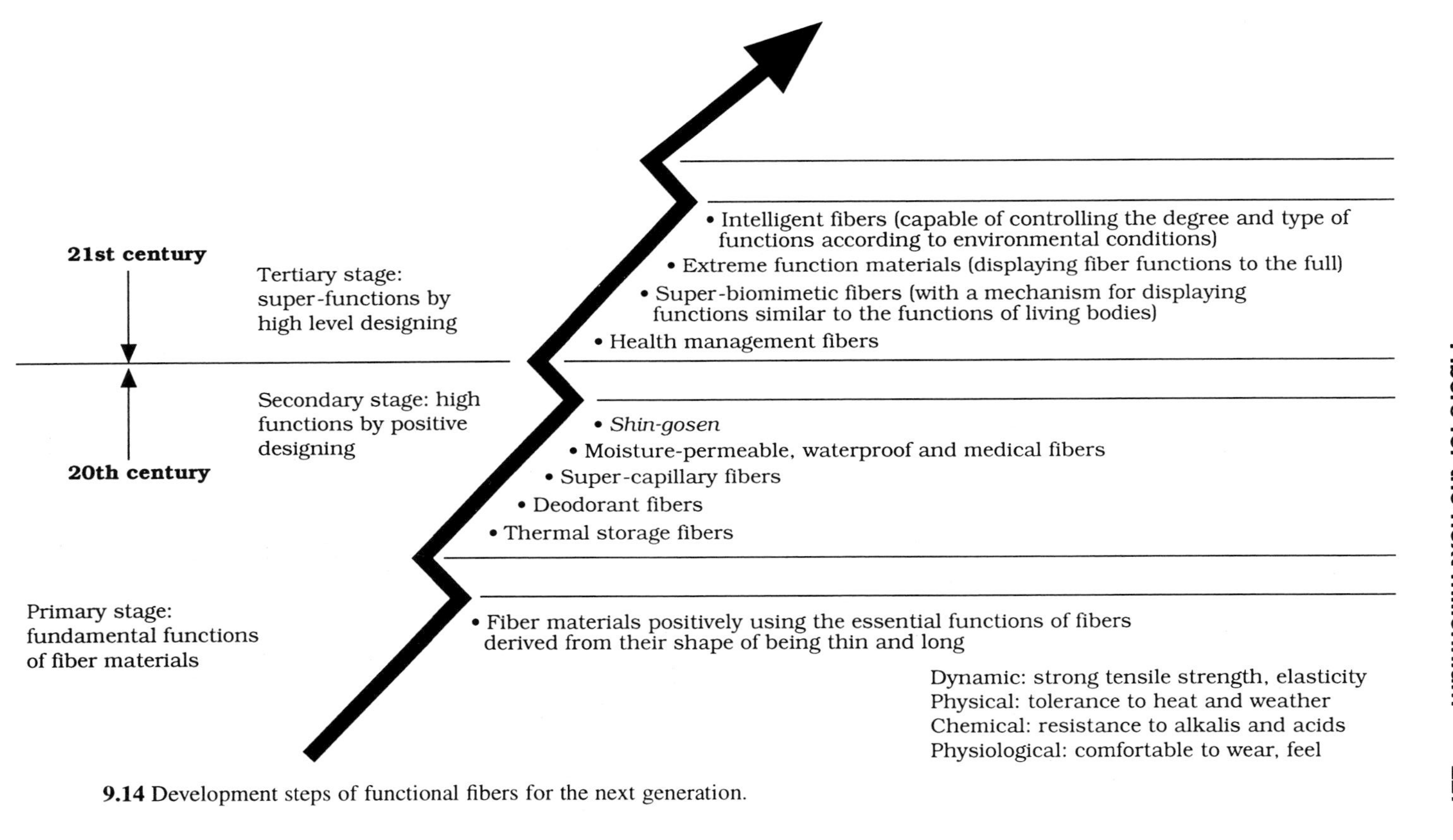

9.14 Development steps of functional fibers for the next generation.

possible impact on the global environment in developing super-function fiber materials, with such performance as strength, elasticity, heat resistance and environment tolerance and having optical, electronic and magnetic properties. In addition, a broad range of problems, including harmony between people and fibers, between communities, fibers and human mental activities, should be addressed. Mind, sensibility, health, animal and plant systems and comfort should be dealt with as the subjects of human engineering. Figure 9.15 gives examples of the research areas and social infrastructure for such *Kansei* (aesthetic) engineering.

9.6 Super-biomimetic fiber materials

Biomimetics, the design of new fiber materials through the study of living systems, has already achieved a great deal (Fig. 9.16). Because of this, it is certain that learning from the system of living bodies and studying their structures and functions will be important to the development of fiber science and technology in the next generation and will provide a fund of new ideas and concepts. "Super-biomimetic fiber materials" can be developed in this way. However, this does not mean the mere copying of living bodies: it is to make clear their structures and functions, to learn from their high-level mechanism of displaying their functions, and to apply the mechanisms to molecular design and materials design. An interdisciplinary approach will be required to achieve this.

9.7 Super-natural materials

Biotechnology has the potential to produce environmentally friendly fibers from polysaccharides, proteins and other materials found in nature. Figure 9.17 illustrates what can be achieved using polysaccharides.

The research and development of synthetic fiber materials which began to develop in the first half of this century is heavily dependent on petroleum. The world's production of raw fibers is increasing mainly as a result of growth in synthetic fibers and the output of chemical fibers is now nearly half of that of all fibers. However, the reliance of chemical industry on petroleum will reach its limit sooner or later.

In view of the rapidly increasing global population, it is very important to promote further the recycling of resources and to strengthen efforts to save energy and other resources. It is also essential to continue research and development on new, environment-friendly fiber materials. Potential fiber materials include, for example, wood-based biomass (including wood cellulose), starch, biodegradable polyester, carbon dioxide, coal and amino

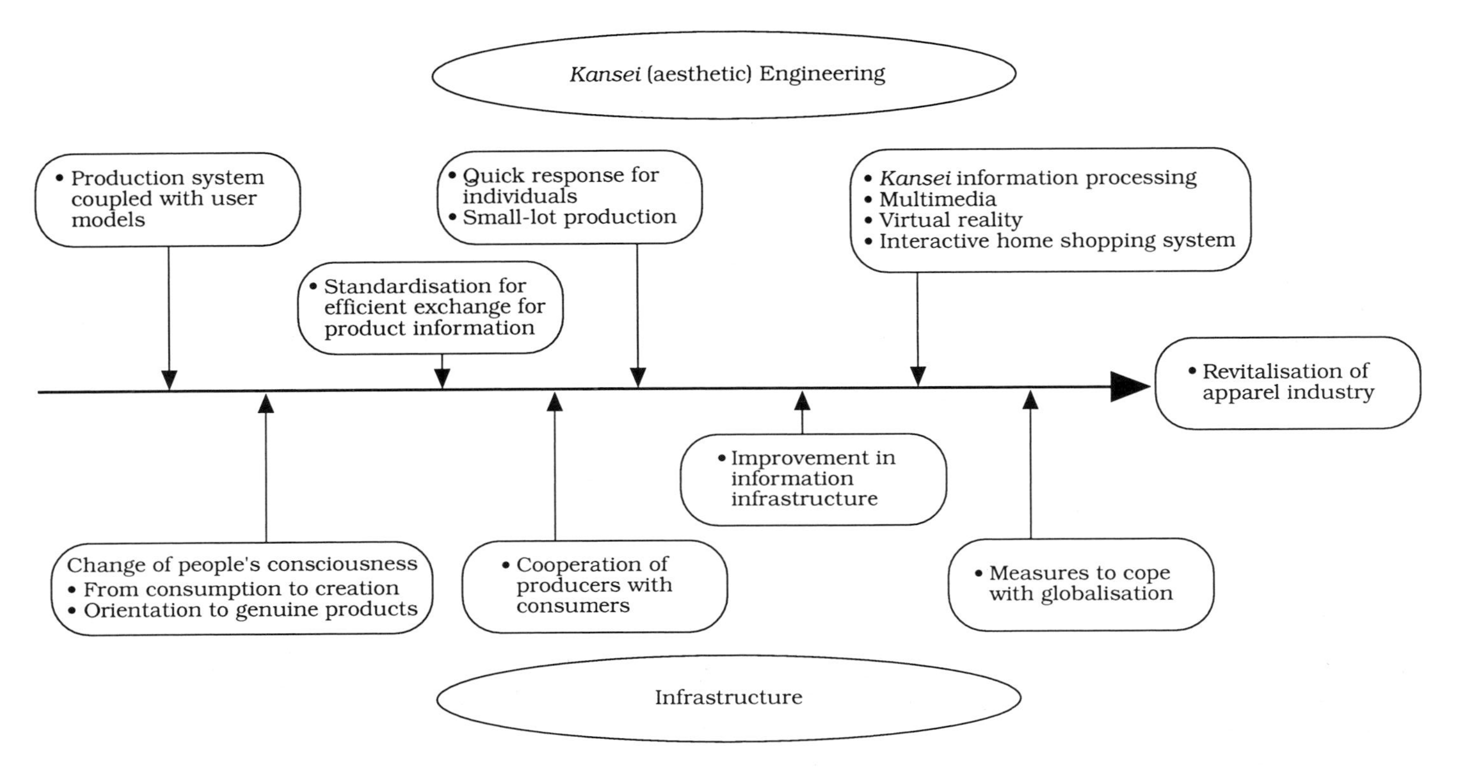

9.15 Research areas and infrastructure of Kansei (aesthetic) engineering.

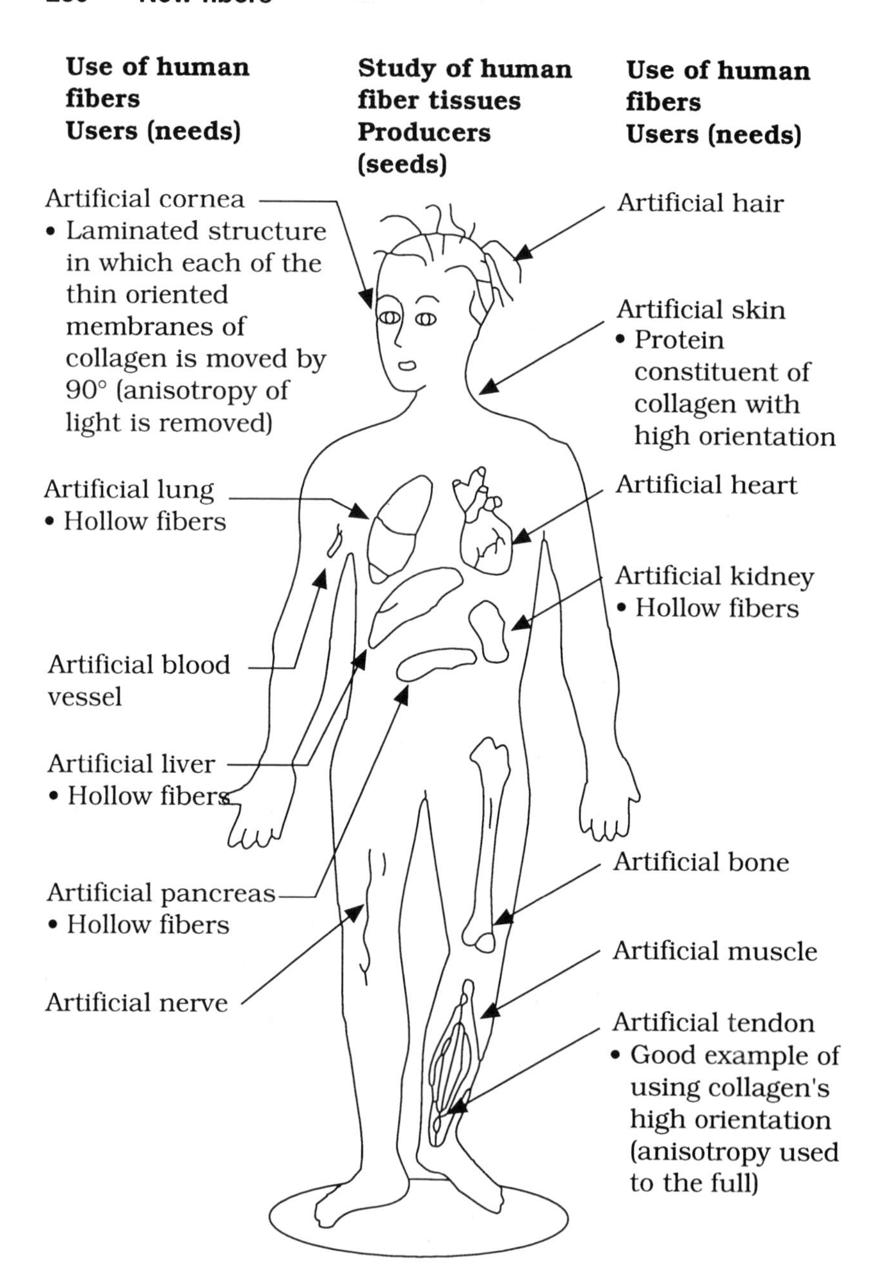

9.16 Designing of materials better than living bodies by learning from them.

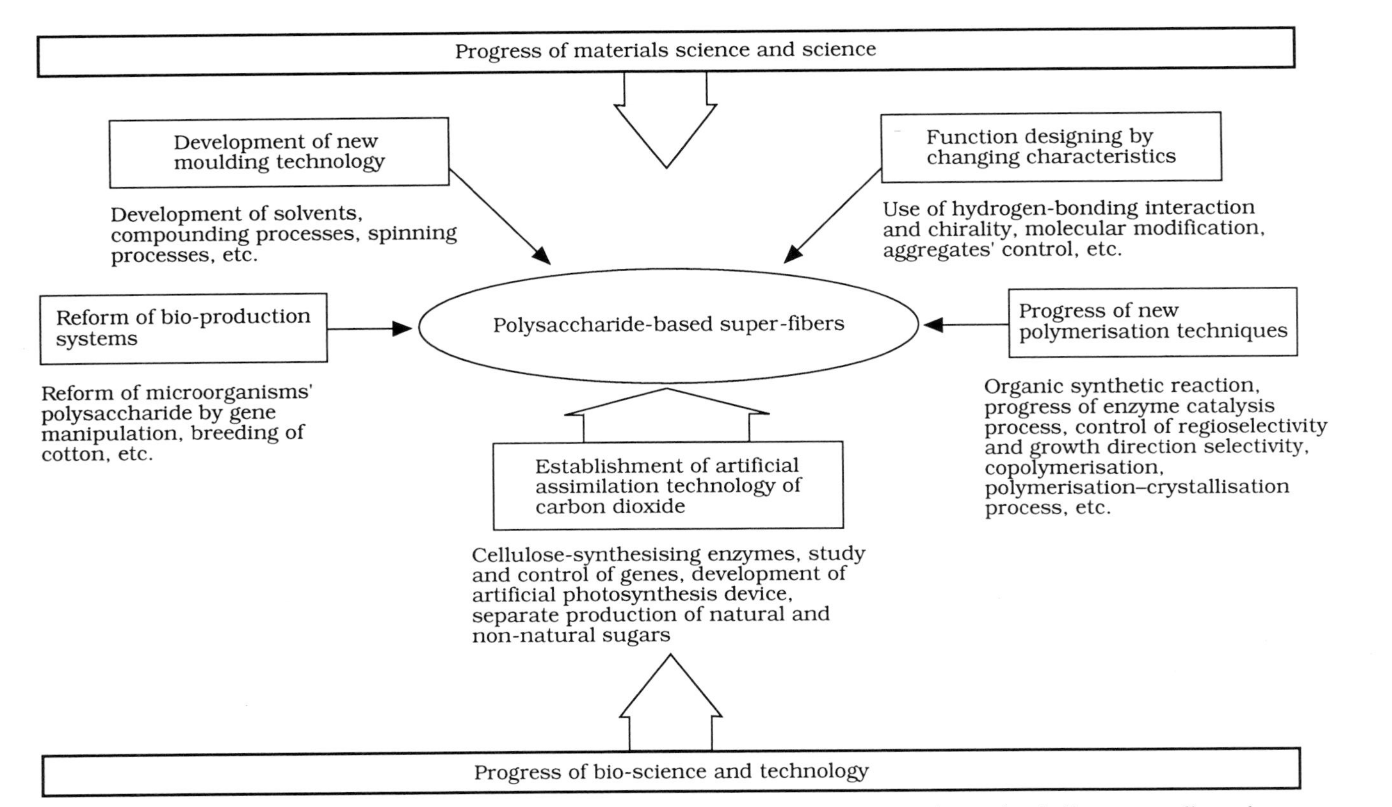

9.17 Process of creating new polysaccharide-based super-fibers: four basic approaches and a challenge to realise a dream.

Table 9.1. Forecast of the development of fibers and fiber science in the next generation

(Year) 1995	Super-functions	Super-performance	Raw materials and production technology
2005	● High-sensitivity fibers ● High-performance antibacteria fibers	● Super-strong fibers substituting for asbestos ● New fibers with multiple layers and diverse structure	● After-processing, workerless production of synthetic fibers ● Designing system to cope with consumer preference to personality
2015	● Environmental change responsive fibers (intelligent fibers; having difference in performance) ● Biodegradable fibers	● Super-fibers in the second generation ● Good cost-performance super-fibers	● Fiber recycling technology (having difference in technology) ● Speedy cultivation of natural fibers ● High-speed spinning of 8,000–10,000 m/min
2025	● Optical fibers (GI-type POF) ● Nerve fibers effectively conveying weak electric currents (artificial nerve) ● Motion function retaining fibers (artificial muscle)		● High-performance plant fiber production technology ● Environment-friendly fiber production technology ● More efficient rayon production process
2035	● Superconductive fibers ● Hazardous substance (NO_x, SO_x) absorbing fibers ● Uranium-absorbing fibers	● Carbon fiber expanded into automobile use	● Synthetic fiber materials made of non-petroleum material by carbon dioxide fixation etc. ● Highly efficient production technology of natural fiber materials by biotechnology

acids as well as inorganic-based substances, such as silicon compounds. In the next generation, human needs will be respected much more than at present, and so in those areas where there is a close relationship to organic substances and living organisms, such as clothes and fiber materials for medical use, human-friendly fibers will be required. Therefore, such naturally occurring materials will have a higher potential in the future.

Examples include the use of wood-based biomass and conversion of starch into fibers by the salting-out process. An important theme for the distant future is to fix carbon dioxide into organic materials. Plants fix it to make cellulose (cotton) and other organic substances: petroleum, coal and other organic substances were originally made by the fixing of carbon dioxide by plants, too. Petroleum generates this gas when it burns and if science and technology make it possible to fix it efficiently and to use it as a fiber material, it will make an immeasurable contribution to humankind. The range of applications will certainly increase and new materials with different properties from those presently available will be needed (Table 9.1).

9.8 Resources recycling

From now on, all industries will be required to maintain the balance between economic growth and environmental protection. The fiber and textile industry is no exception. Greater efforts to reduce and reuse industrial wastes are needed from the viewpoint of environmental protection and energy and resources conservation (Fig. 9.18).

The recycling of resources can be broadly divided into thermal, material and chemical sectors. In the fiber and textile industry, thermal recycling is intended to recover heat energy generated from the incineration of fiber wastes as thermal or electrical energy. This method, although easily practicable, does not mean the recycling of resources. Material recycling recovers polymers from fibers or plastics, and at present, the idea of transforming polyethylene terephthalate (PET) into fibers is most economical and widely used for practical purposes. But there is concern about this method which is apt to let impurities mix into recovered polymers, resulting in declined quality and spinning stability. Chemical recycling recovers monomers from waste fibers by polymer decomposition. This is the method of the future. Impurities can be easily removed from recovered monomers, so their quality will be made exactly equal to virgin monomers. An important consideration in all three sectors is to establish an economical collecting system and an efficient recovery technology and to develop commodities using recovered materials. The key point of material and chemical recycling in particular is how to collect and separate wastes. In this context, it may be

1 Waste handling

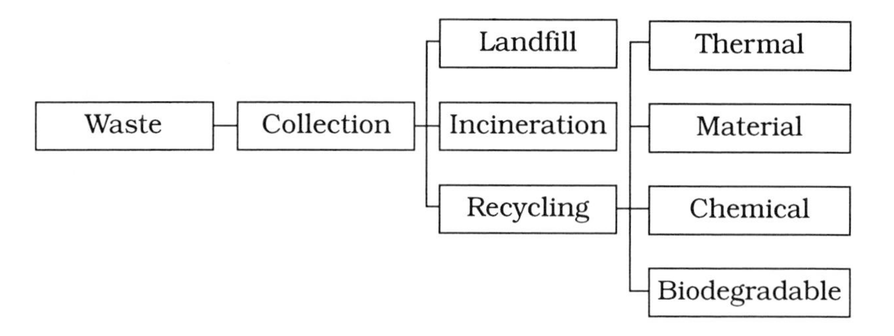

2 Comparison of recycling methods

Method	Sorting	Applied for	Remarks
Thermal	Not required	Energy recovery • Electric power • Local heating	Efficient recovery system
Material	Required	Fiber or plastics	Proper applications
Chemical	Required	Any products	Economical recovery technologies

3 Future Works

- Economical collecting system from consumers
- Efficient recovery technologies
- Appropriate application development

9.18 Recycling: environment protection, energy saving and resource saving.

argued that the development of those products that can be easily recycled will be an important task to be carried out in the years ahead.

9.9 Fibers for health

The presence of fiber in food is beneficial to health. The subject of such benefits is rapidly expanding, and is worthy of a place in the future application of the fiber concept. Dietary fiber (DF) was originally described as the skeletal remains of plant cells in the diet that are resistant to hydrolysis by the digestive enzymes of humans. This excluded polysaccharides in the diet present as food additives (i.e. exudate gums, modified celluloses and starches), so the definition was expanded to include all polysaccharides and lignin that are not digested by the endogenous secretions of the human digestive tract. Dietary fiber thus mainly comprises non-starch polysaccharides (NSPs) but owing to imprecise methods of analysis it also includes oligosaccharides, polyphenolics (includes lignin), cutin, waxes, suberin, phenolic esters and inorganic constituents. These materials are very diverse, having many different nutritive and physiological properties. The levels of the individual components are also very difficult to measure. This has resulted in DF being defined by some workers, particularly Englyst, as the "polysaccharides which are resistant to the endogenous enzymes of man" (British Nutrition Foundation, 1990). Despite these variations and possible future changes, it is certain that the polysaccharides will remain central to the concept of DF. Such dietary fibre polysaccharides are here classified (Table 9.2).

DF is not digested by the endogenous secretions of the gastrointestinal tract, but undergoes anaerobic fermentation in the colon. The products are non-utilisable gases (for example, hydrogen, methane and carbon dioxide) and organic acids, particularly short-chain fatty acids. These can be utilised in human metabolism after absorption from the intestine.

Health professionals and medical groups now recognise that foods rich in dietary fiber have significant and beneficial effects on human health, mainly as a result of the bulking action in the colon and the products formed there by fermentation. The American Dietetic Association specifically recommends eating 20–35 g of fiber every day. Epidemiological research supports the theory that high intakes of dietary fiber protect against degenerative diseases. High-fiber diets reduce disease risk by increasing faecal bulking, decreased transit time, reduced blood cholesterol levels and assisting in the control of blood sugar levels. It is the "soluble" fibers (more correctly hydrocolloids") that are mainly responsible for reducing blood cholesterol levels, and helping control the rise in blood glucose levels following a meal, thereby reducing insulin requirements in some people with diabetes. The fatty acids, such as propionic acid, are metabolised by the liver, where it inhibits cholesterol synthesis. There is a great deal of evidence too for dietary fiber's role in cancer prevention. The USA National Research Council found colon cancer negatively correlated with high dietary fiber intake. Finally, DF, because it is not broken down in the intestine, has a low calorific value, so contributing to health without adding to obesity.

Table 9.2. Dietary Fiber Polysaccharides

Main source	Soluble in water*	Major groups	Components	Polysaccharide type	Foods
Storage material	Partly	Starch	Amylose Amylopectin	α(1-4)-glucan α(1-4,1-6)-glucan	Fruits, seeds tubers
Structural material	No	Cellulose		β(1-4)-glucan	All cell walls
		Non-cellulosic			
	Yes		Pectic substances	Galacturonans Arabino-galactans	Mainly fruits and vegetables
			Hemi-cellulose		Cereals
	Slightly			Arabino-xylans Glucurono-arabinoxylans β-glucans	Cereals Fruits and vegetables Cereals
Non-structural materials	Yes Yes	Mucilages Gums		Diverse and complex hetero-polysaccharides	Algal seaweed exudates, seeds and fruits

In the next millennium there is a great opportunity to link fiber science and technology with the fabrication of new types of DF, which can function as a result not only of their metabolic behaviour, but also of their physical performance in various organs of the human body.

9.10 Conclusion

The scope of fiber science is very broad. The future of fiber research will be directed to produce only what is needed. Only innovative products will be able to open up new markets and new horizons for the textile industry. Table 9.1 shows a forecast of the development of fibers and fiber science into the new millennium.

As a famous Scandinavian economist said, "It is very dangerous to make predictions, particularly about the future!" It is in this spirit that the predictions are made. New science, not yet developed, can produce products beyond our present perception. This has been the way in the past.

We can give as typical examples the development of *Shin-gosen*, super-fibers, optical fibers and fiber-reinforced plastics, which did not exist at all 30 years ago. In the area of fibers and their surrounding areas, it is expected that there will emerge via innovative science and technology materials that will surpass these fibers.

To realise this, it is essential to invest in future research and researchers. In particular, the education of software researchers will be important, since demand will increase for human-friendly fibers. Fiber science 20 or 30 years from now will depend on development by humans.

What is needed is not simply the conveyance of knowledge but the development of truly creative researchers. The textile industry is required to shift its emphasis from "quantity" to "quality" as the 21st century dawns. The industry must adapt itself to the dynamism of the market economy.

Further reading

Chapter 1 Birth of the new fibers

1. *High Technology Fibers: Key to Stronger High Quality Products*, Technical Insights, Inc., Fort Lee, New Jersey, 1985.
2. *High Technology Fibers: Handbook of Fiber Science and Technology*, Vol. III, Ed. by M. Lewin and J. Preston, Marcel Dekker, New York, 1985.

Chapter 2 The super-fiber with new performance

2.1 Two streams of super-fiber

1. *ISF '85 Proceedings*, Society of Fiber Science and Technology, Japan, 1985.
2. S. L. Kwolek, US Patent 3, 671, 542.
3. P. Smith and P. J. Lemstra, *J. Mater. Sci.*, **15**, 505 (1980).

2.2 The quest for a strong fiber

1. *Developments in Oriented Polymers–2*, Ed. by I. M. Ward, Elsevier Applied Science, London, 1987.

2.3 From "shish kebab" to "gel-spinning"

1. A. Peterlin, *J. Polym. Sci.*, **A-2**, 1151 (1969).
2. A. Keller and F. M. Willmouth, *J. Macromol. Sci. (Phys.)*, **B6**, 493 (1972).
3. P. J. Lemstra, R. Kirschbaum, T. Ohta and H. Yasuda, in *Developments in Oriented Polymers–2*, Ed. by I. M. Ward, pp. 39–77, Elsevier Applied Science, London, 1987.

2.4 The aramid fiber race in Europe, the USA and Japan

1. *Kevlar 149*, Du Pont Brochure, June, 1987.
2. *Data Manual for Kevlar 49 Aramid,* Du Pont, May, 1986.
3. *Kevlar Fiber Optics Data Base*, Du Pont, June 1987.
4. *Twaron Technical Documentation*, AKZO, 1985.

5. *Twaron: The High Performance Aramid Fiber*, AKZO, 1986.
6. T. Takada, Aramid fiber, in seminar on A Spronc (Japan), May 1987.
7. H. Imuro and N. Yoshida, *Chemiefasern Textilindustrie*, **37/89**, T4 (1987).

2.5 Polyacetal fiber

1. T. Komatsu, S. Enoki and A. Aoshima, *Polym. Preprints (Jpn)*, **35**, 3712 (1986).
2. T. Komatsu, S. Enoki and A. Aoshima, *Polym. Preprints (Jpn)*, **36**, E666 (1987).

2.6 Strong Vinylon RM

1. Kuraray, British Patent 1,319,099.
2. Kuraray, British Patent 2,073,653.
3. Kuraray, European Patent 99,053.
4. Kuraray, US Patent 4,524,042.

2.7 New liquid crystalline polymers: engineering plastics for the next generation

1. G. W. Calundann, US Patents 4,067,852; 4,161,470; 4,185,996; 4,256,624.
2. W. J. Jackson, Jr and H. F. Kuhfuss, US Patent 3,778,410.
3. L. F. Charbonneau and G. W. Calundann, US Patent 4,330,457.

2.8 Vectran – a fully aromatic polyester fiber

1. Tai-Shung Chung, *Polym. Eng. Sci.,* **26**, 901, (1986).

2.9 Developing polyallylate fiber

1. T. Kunugi, T. Ohta and K. Yabuki, *New Polymer Materials 9*, Kyoritsu Shuppan, Tokyo, 1988.

2.10 The ACM industy in the USA

1. A. S. Wood and R. Juran; *Modern Plastics*, 47–53, July (1985).

2.12 Final stage of pitch-based carbon fiber development

1. P. Ehrburger and J.-B. Donnet, in: *High Technology Fibers: Handbook of Fiber Science and Technology*, Vol. III, Ed. by M. Lewin and J. Preston, pp. 169–220, Marcel Dekker, NewYork, 1985.
2. *Carbon Fibers: Technology, Uses and Properties*, Plastic and Rubber Institute, London, 1986.
3. K. Okuda, *New Polymer Materials 13*, Kyoritsu Shuppan, Tokyo, 1988.
4. H. Blumberg, *Fibers for Composites – Status Quo and Trends*, 35th International Man-Made Fibers Congress, Dornbirn, Austria, 1996.
5. V. Meier: Construction and Building Materials, **9**(6), pp. 341 (1995).

Chapter 3 High-touch fibers

3.2 Challenge of ultra-fine fibers

1. Toray, Japanese Patent 61-68573.
2. TorayY, US Patent 4,350,006.

3.4 Chameleonic fabrics

1. Toray, *Sportstyle*, 52–53 (1988).
2. Toray, US Patent 4,784,474.

3.7 Perfumed pantyhose

1. Ichikawa *et al.*, *Dyeing Finishing*, **7**, 168 (1989).
2. Kanebo, US Patent 4,882,220.

3.8 Power fibers that store solar energy

1. Unitika, US Patent 4,636,424.
2. S. Kagechi and M. Fujimoto, in: *Proceedings of '88 Cold Region Technology Conference*, 1988.

3.9 Iridescent textiles

1. K. Matsumoto, T. Okada, N. Tsutsumi and T. Kiyotsukuri, *J. Text. Machine Soc. Jpn*, **42**, T19 (1989).
2. K. Matsumoto, Y. Kitagawa, N. Tsutsumi and T. Kiyotsukuri, *J. Text. Machine Soc. Jpn.*, **42**, T160 (1989).

Chapter 4 Biomimetic chemistry and fibers

4.1 Applications of morphology/structure

1. M. Okamoto, *JTN*, **276**, 94 (1977).
2. M. Okamoto, *JTN*, **278**, 77 (1978).
3. US Patent, 4,476,186.
4. Toray, *Textile Horizons*, 30–31 (1985).
5. F. Shibata and S. Kawasaki, *Sen-i-Gakkaishi*, **44**, 94 (1988).
6. Kuraray, US Patent 4,514,459.
7. Kuraray, European Patent 0,127,891.
8. Kuraray, US Patent 4,254,182.
9. Kuraray British Patent 2,016,364.
10. H. Shirai, *Fragrance J.*, **86**, 75 (1987).
11. H. Shirai, *J. Odor Res. Eng.*, **20**, 36 (1989).

4.2 Hybridisation technology

1. T. Akaike, S. Tobe, A. Maeda, A. Yagawa, H. Nomoto, K. Kobayashi and

H. Sumitomo in: *Proceedings of the 3rd Japanese–French Biomedical Technologies Symposium*, 1989.
2. J. C. Bigelow, J. C. Giddings, Y. Nabeshima, T. Tsuruta, K. Kataoka, T. Okano, N. Yui and Y. Sakurai *J. Immunol. Methods,* **117**, 289 (1989).
3. K. Kataoka, T. Okano, Y. Sakurai, A. Maruyama and T. Tsuruta, in: *Multiphase Biomedical Materials*, Ed. by T. Tsuruta and A. Nakajima, pp. 1–19, VNU, Utrecht, 1989.
4. A. Maruyama, T. Tsuruta, K. Kataoka and Y. Sakurai, *Biomaterials*, **10**, 393 (1989).

Chapter 5 Biopolymer frontiers

5.1 Mimicking the functions of enzymes and coenzymes

1. O. Hirabaru, T. Nakase, K. Hanabusa, H. Shirai, A. Takemoto and N. Hojo, *J. Chem. Soc., Dalton Trans.*, 1485 (1984).
2. T. Matsunaga, R. Tomoda, T. Nakalima, N. Nakamura and T. Komie, *Appl. Envir. Microbiol.*, **54**, 1330 (1988).

5.2 Polysaccharides in semiconductors and medicine

1. M. Endo, M. Sasago, Y. Hirai, K. Ogami and T. Ishihara "Proc. SPLE Santa Clara, USA", *Laser in Microlithography*, **774**, 138 (1987).
2. G. Chthara, J. Hamuro, Y. Maeda, Y. Arai and F. Fukuoka, *Nature*, **225**, 943 (1970).
3. T. Norisue, T. Yanaki and H. Fusita, *J. Polym. Sci., Polym. Phys.*, **18**, 547 (1970).
4. T. Norisue, Macromol. *Chem. Suppl.*, **14**, 105 (1985).
5. H. Furue, *Drugs Today,* **23**, 335 (1987).
6. T. Fujita, Y. Yasuda, Y. Takakura and M. Hoshida, *J. Contr. Release*, **11**, 149 (1990).
7. J. Sunamoto and T. Sato, in: *Multiphase Biomedical Materials*, Ed. by T. Tsuruta and A. Nakajima, pp. 167–190, VNU, Utrecht, 1989.
8. J. Sunamoto, in: *Cellulosics Utilisation (Proceedings Nisshinbo International Conference)* Ed. bv H. Inagaki and G. O. Phillips, pp. 246–254, Elsevier, London, 1989.
9. J. Kisugi, H. Kamiya and M. Yamazaki, *Cancer Res.*, **47**, 5649, (1987).
10. M. Yamazaki, K. Kimura, J. Kisugi, K. Maramoto and H. Kamiya, *Cancer Res.*, **49**, 3834 (1989).

5.3 Biomass of crab and shrimp shells

1. S. Tokura, S. Nishimura, N. Nishi, K. Nakamura, O. Hasegawa, H. Sashiwa and H. Seo, *Sen-i-Gakkaishi,* **48**, 288 (1987).
2. S. Tokura, J. Yoshida, N. Nishi and T. Hiraoka, *Polym. J.*, **14**, 527 (1982).
3. S. Kishimoto and K. Tamaki, *Acta Dermatol., Kyoto*, **82**, 471 (1987).
4. H. Seo and Y. Kinemura, in: *Chitin and Chitosan (Proc. 4th Int. Conf. Chitin and Chitosan)*, Ed. by G. Skyjak-Braek, T. Anthonsen and P. Sandford, pp. 585–588, Elsevier, London, 1989.

5.4 New applications of silk

1. C. Takabayashi, *Sen-i-Gakkaishi*, **45**, 268 (1989).
2. S. Mizushima, Japan Patent 63-64550.

5.5 Fibers produced by bacteria

1. S. Yamanaka, K. Watanabe, N. Kitamura, M. Iguchi, S. Mitzuhashi, Y. Nishi and M. Uryu, *J. Mater. Sci.*, **24**, 3141 (1989).
2. Y. Doi, Y. Kanesawa, M. Kunioka and T. Saito, *Macromolecules*, **23**, 26 (1990).

5.6 New functions for cellulose

1. H. Hayatsu, in: *Advances in Mutagenesis Research – 1*, Ed. by G. Obe, pp. 1–26, Springer, Heidelberg, 1990.
2. US Patent 4,634,470.
3. K. Sudo, K. Shimizu and K. Sakurai, *Holzforschung*, **39**, 281 (1985).
4. G. O. Phillips, "Rediscovering cellulose", *Chemistry in Britain*, **October**, 1006–1009, 1989.

5.7 Utilisation of protein functionality

1. US Patent 4,223,984.
2. US Patent 4,260,228.
3. H. Takenaka, K. Fujioka, S. Nakaso, K. Nakaya, Y. Nakamura and Y. Kurihara, *J. Biol. Chem.*, **264**, 6655 (1989).
4. N. Takahashi, H. Hitotsuya, H. Hanzawa, Y. Arata and Y. Kurihara, *J. Biol. Chem.*, **265**, 134–143 (1990).
5. K. Kurihara M. Kashiwayanagi, K. Yoshti and Y. Kurihara, *Comments Agric. Food Chem.*, **2**, 1 (1989).

Chapter 6 Progression of high-tech fibers

6.1 Utilisation of unused resources

1. N. Tsujimoto, *Japan Tappi J.*, **43**, 167 (1989).
2. N. Shiraishi, in: *Cellulosics Utilisation (Proc. Nisshinbo Intern. Conf.)* Ed. by H. Inagaki and G. O. Phillips, pp. 97–109, Elsevier, London, 1989.

6.2 Biotechnology and fibers

1. Y. Ikada, in: *High Technology Fibers*, Ed. by M. Lewin and J. Preston, Marcel Dekker, New York, 1989.
2. H. Iwata, H. Amemiya, T. Matsuda, H. Takano and T. Akutsu, *J. Bioactiv. Compatible Polym.* **3**, 356 (1988).
3. T. Akizawa, T. Kitaoka, S. Koshikawa, T. Watanabe, K. Imamura, T. Tsurumi, Y. Suma and S. Eiga, *Trans. Am. Soc. Artif. Intern. Organs*, **32**, 76 (1986).
4. J. P. Mortensen *et al.*, *Trans. Am. Soc. Artif. Intern. Organs*, **33**, 570 (1989).
5. A. R. Katz *et al.*, *Surgery, Gynaecology Obstetrics*, **161**, 213 (1985).

6. T. Tagawa *et al.*, *J. Polym. Sci., Polym. Phys.*, **18**, 971 (1980).
7. M. Kodama, T. Sakai and K. Tsuda, *Jpn. J. Artif. Organs*, **17**, 647 (1988).
8. H. Ichijo, K. Ichimura, H. Uedaira, T. Suehiro, J. Nagasawa, A. Yamauchi and N. Aisaka, in: *Polymer Gels: Fundamentals and Biomedical Applications*, Ed. by Y. Osada, K. Kajiwara, A. Yamauchi and D. De Rossi, Plenum, London, 1990.
9. S. Manabe, T. Tsurumi, G. Ishikawa, M. Satomi, T. Yamashiki, Y. Hamamoto, K. Yamaguchi, S. Kobayashi and N. Yamamoto, *Membrane*, **14**, 77 (1989).
10. S. Sekiguchi, K. Ito, M. Kobayashi, H. Ikeda, T. Tsurumi, G. Ishikawa, S. Manabe, M. Satomi and T. Yamashiki, *Membrane*, **14**, 253 (1989).

6.3 Electronics and fibers

1. T. Fukumoto, M. Kodima and A. Nagae, *9th LCCCS Proceedings*, pp. 484–488 (1988).
2. Japan Patent 61-55204.

6.5 Fibers in space

1. S. Kibe, Y. Fujimori, K. Kawakami, T. Hamaguchi, M. Tomita, M. Iwata and Y. Kumagiri, in: *Proc. 15th Intern. Symp. Space Tech. & Sci.*, Tokyo, 1986.
2. F. H. Greenwood and R. J. Balinskas, *SAE Technical Paper Series 85*, 1333 (1985).
3. H. C. Wright, III, *SAE Technical Paper Series 85*, 1335 (1985).

6.6 Fibers and nuclear power

1. T. Seguchi, *Sen-i-Gakkafshi*, **44**, 336 (1988).

6.8 Fibers for geotextiles

1. D. J. Hannant, *Fiber Cements and Fiber Concretes*, John Wiley & Sons, New York, 1978.
2. Y. Tanno, T. Yano, Y. Iwasa, H. Kihara and K. Matsunaga, *Proc. IASS-MSU Symp.*, pp 631–640, Istanbul, 1988.

6.9 Fibers in the ocean

1. *D/W 26,000 MT Modern Sail-Assisted Log and Bulk Carrier, Usuki Pioneer,* Technical Information Sheet, by JSIF, 1985.
2. E. E. Johanson, *Proc. 7th World Dredging Conf.*, pp. 183–206 (1978).
3. K. Ohno, Y. Masuda, T. Miyazaki, K. Takahashi, M. Ukita and T. Tuzuku, *JAMSTECTR*, **3**, 50 (1979).
4. D. W. Skelly, *Oceans (USA)*, **1**, 600 (1985).
5. T. Kano, H. Minami and H. Kawamoto, *Proc. 6th Symp. Coastal & Ocean Management*, South Carolina, 1989.
6. N. W. M. John, *Geotextiles*, Blackie, Glasgow & London, 1988.
7. W. T. Bakker, J. Bax, D. Grootenboer and W. H. Tutuarima, *La Houille Blanche*, **8**, 661 (1978).

Chapter 7 New High-tech fibers and *Shin-gosen*

7.1 Various categories of high-tech fibers

1. M. Lewin and J. Breton, *High Technology Fibers. Part A*, Marcel Dekker, Inc., New York, 1985.
2. Tatsuya Hongu, *Sen-i Gakkaishi*, **49**, 324 (1993).
3. Tatsuya Hongu, *Digest of Japanese Industry & Technology (DJIT)*, No. 297, p. 17, Japan Trade & Industry Publicity, Inc., Tokyo, 1996.
4. Takeaki Miyamoto and Tatsuya Hongu, *New Fiber Materials*, p. 128, Nikkan-kogyo Shimbun-sha, Tokyo, 1992.

7.2 Development of Shin-gosen

1. Takeshi Nishita, *The 23rd Summer Seminar Proceedings*, p. 99, The Society of Fiber Science & Technology, Japan, 1991.
2. *Report of special investigation "High-tech High-touch fiber"*, **3**, No. 122, Osaka Chemical Marketing Center, Osaka, Kansai, 1992,
3. Tatsuya Hongu, "Everythingyou should know about *Shin-gosen*", in the special issue of *Engineering Materials*, p. 11, Nikkan-kogyo Shimbun-sha, Tokyo, 1994
4. Public Liaison Committee, Commemoration of 50th Anniversary, gathered information by questionnaire about *Shin-gosen*, *Sen-i Gakkaishi*, **51**, 136 (1995).
5. Masao Matsui, *Advanced Fiber Spinning Technology*, p. 115, Woodhead Publishing Limited, Cambridge, 1994.
6. Tatsuya Hongu, *Trigger*, p. 113, Nikkan-kogyo Shimbun-sha, Tokyo, 1993.
7. Keitaro Nabeshima, *Recent Review of Textile Technology Proceedings*, p. 6, Society of Fiber Science & Technology, Japan, 1994.
8. Tatsuya Hongu, *Trigger*, p. 90, Nikkan-kogyo Shimbun-sha, Tokyo, 1994.

7.3 Design of specialist fibers

1. Tetsunori Higuchi, *The 27th Summer Seminar Proceedings*, p. 46, Society of Fiber Science & Technology, Japan, 1996.
2. Miyoshi Okamoto, *Advance Fiber spinning Technology*, p. 187, Woodhead Publishing Ltd., Cambridge, 1994.
3. Masao Matsui, *Kako Gijutsu*, **22**, No. 1–2, 1987.

7.4 Fabrics for relaxation using 1/f fluctuations

1. T. Musha, *Yuragi no sekai (World of fluctuations)*, Kodansha, Japan, 1980.
2. T. Musha, *Oyo Butsuri*, **54**, 5(1985).
3. T. Musha, K. Takakura, and J. Ikebe, *Yuragi no Igaku-1/f Yuragi Kenkohou*, Shujun-sha, Japan, 1985.
4. T. Musha and Y. Sawada, *Yuragi, Kaosu, Furakutaru*, Nihon Hyoron-sha, Tokyo, 1991.
5. T. Suzuki, *1/f Yuragi, Riron Oyo Shohin no kanosei*, Nikkei Select Products, **6** (1993).
6. T. Musha, *Yuragi no Hasso*, NHK Shuppan, Japan, 1994.
7. T. Musha, *Nippon Onkyo Gakkaishi*, **50**, 6 (1994).

Chapter 8 Cellulosic man-made fibers

8.1 New solvent systems

1. Patents for Lyocell: US Patents 4246221; 819080; 819081; 189082.
2. I. P. O'Brian: US Patents 4464323; 4501884 (assigned to Du Pont).
3. W. Rimpp, "Marketing of NewCell filament yarns", *Chemical Fibers International*, **46**, 22 (1996).
4. V. Eichinger and M. Eibl: Lenzing lyocell – an interesting cellulose fiber for the textile industry, *Chemical Fibers International*, **46**, 26 (1996).

8.2 New cellulosic fiber derivatives

1. H. Struszczyk in *Cellulose and Cellulose Derivatives: Physical — Chemical Aspects and Industrial Applications* Ed. by J. I. Kennedy, G. O. Phillips and P. A. Williams, Woodhead Publishers Ltd, Cambridge, 1995.
2. *Biological Transformation of Wood,* Ed. by W. Liese, Springer Verlag, Berlin, 1985.

8.3 New environmental and cost saving developments

1. J. Schmidtbauer, "An environmentally-sound technology for the production of total chlorine-free fibers", in: *Challenges in Cellulosic Man-Made Fibers*, Viscose Chemistry Seminar, Akzo Nobel, Stockholm, 1994.

8.4 Life-cycle assessment

1. *Eco-profiles of the European Plastic Industry*, a report for the European Centre for Plastics in the Environment, Brussels, 1993.
2. Society of Environmental Toxicology and Chemistry, *Guidelines for Life Cycle Assessment: A Code of Practice*, 1993.

8.5 Cellulose: the renewable resource

1. Kenji Kamide, "Cellulose and cellulose derivatives: recent advances in physical chemistry", *Adv. Polym. Sci.*, **83**, 1–56 (1988), and papers quoted therein.

Chapter 9 Fibers for the next millennium

1. K. Yabuki, "Wonder Fiber: Is PBO the revolutionary new fiber for the 21st century", *Look Japan*, 24–25 (1995).
2. H. H. Yang, *Kevlar Aramid Fiber*, John Wiley, 1993.
3. A. Buckley and G. A. Serad, *Encyclopedia of Polymer Science and Engineering*, Vol. 11, 2nd Edition, p. 601, 1988.
4. S. Kumar, in: *Intern. Encyclopaedia of Composite*, Vol. 4, p. 51, Ed. by S. M. Lee, VCH, New York, 1991.
5. K. Yabuki, *Sen-i Gakkaishi*, **48**, 448 (1995)
6. K. Yabuki, "Poly (*p*-phenylenebenzobisoxazole) Fiber", *The Twelfth Annual Meeting of the Polymer Processing Society*, Sorrento, Italy (May 27–31), 1996.

9.3 The next stage: technological improvements

1. *New Fiber Science: Challenge to New Frontiers*, Ed. by The Committee for Research of Fiber Science for the Next Generation, Publishing Department, International Trade and Industry Research Council (Japan), 1995.

9.4 The next century: Respect for people's quality of life and harmony with nature to 9.7 Super-natural materials

1. T. Hongu, "Everything you should know about *Shingosen*", *Eng. Mater., Nikkankogyo-Shimbun-Sha*, 11 (1994).

9.8 Resources recycling

1. T. Miyamoto and T. Hongu, *Introduction to Shingosen*, p. 15, Nikkankogyo-Shimbun-Sha, 1992.
2. *New Fiber Science*, Ed. by Shinohara, S. Ahirai and Konda, p. 132, Baifukan Publishers, 1990.
3. Y. Shimizu, *Fashion Technology of the 21st Century*, p. 1, 3rd Asian Textile Conference, 1996.

9.9 Fibers for Health

1. British Nutrition Foundation, *Complex Carbohydrates in Food*, 1990.
2. H. J. Cumming, "Dietary fiber", *British Med. Bull.*, **37**, 65–70 (1981).
3. G. Livesey, "The energy value of dietary fiber", *Nutritional Rev.*, **5** (1992).
4. *Workshop on the Chemistry and Nutritional Effects of Dietary Fiber*, Ed by S. Samman and G. Annison, CSIRO (Australia), 1991.
5. H. C. Trowell, Ischaemic Heart Disease and Dietary Fiber, *Amer. J. Clinical Nutrition*, **41**, 1103 (1985).

Author index

Trade name index

Company index

Subject index